Intermediate 2 PHYSICS

SECOND EDITION

Andrew McCormick and Arthur Baillie

Editor: Rothwell Glen

Cover Photograph: the cover photograph shows a bundle of optical fibres conducting light. These fibres are made of very pure glass which is coated so as to allow light to be transmitted without absorption or loss of energy. This allows data such as telephone conversations, television pictures and computer transmissions to be sent in large volumes without loss of quality.

HODDER GIBSON
AN HACHETTE UK COMPANY

The Publishers would like to thank the following for permission to reproduce copyright material:

Photo credits
Figure 1.3 © sciencephotos/Alamy; Figure 3.5 © Alex Bartel/SPL; Figure 3.6 © Pasieka/SPL; Figure 4.19 © Curry/ Martyn F Chillmaid; Figure 4.20 © Courtesy of Apple; Figure 5.5 © Adam Hart-Davis/SPL; Figure 6.2 © picturesbyrob/Alamy; Figure 6.3 © Sergio Dorantes/Corbis; Figure 6.4 © Rosenfeld Images Ltd/SPL; Figure 6.5 © AFP/Getty; Figure 6.11 © Martyn F Chillmaid; Figure 6.19 © Hodder Education; Figure 6.23 © NASA/Charles M Duke Jr.; Figure 7.10 © Tibor Bognar/Alamy; Figure 8.2 © Paul Glendell/Alamy; Figure 8.3 © Wellcome Photo Library; Figure 9.1 © Smiley N. Pool/Dallas Morning News/Corbis; Figure 9.8 © Stringer/Malaysia/Reuters/Corbis; Figure 9.9 © Alexis Rosenfeld/SPL; Figure 9.10 © Chad Ehlers/Alamy; Figure 9.11 © Larry Mulvehill/SPL; Figure 9.12 © Image rights reserved by GN Netcom; Figure 10.2 © BL Images Ltd/Alamy; Figure 10.3 © Martyn F Chillmaid; Figure 10.7 © Will and Deni McIntyre/SPL; Figure 10.11 © David Hoffman Photo Library/Alamy; Figure 10.12 © Ted Kinsman/SPL; Figure 10.13 © Martyn F. Chillmaid; Figure 10.17 © Alex Bartel/SPL; Figure 11.11 © Specsavers Opticians; Figure 11.14 © DK/Alamy; Figure 11.16 © Phototake Inc./Alamy; Figure 12.5 © Martin Dohrn/SPL; Figure 12.7 © Josh Sher/SPL; Figure 14.1 © Marty F Chillmaid/SPL; Figure 14.3 © Picture Contact/Alamy; Figure 14.4 © Medical-on-line/Alamy; Figure 14.5 © Josh Sher/SPL; Figure 14.7 © David Parker/SPL; Figure 15.1 © Martin Bond/SPL; Figure 15.2 © Novosti/SPL; Figure 15.6 © David Parker/SPL; Figure 15.7 © Martin Bond/SPL.

Acknowledgements
Relationships and Data Sheet reprinted by permission of the Scottish Qualifications Authority.

Orders: please contact Bookpoint Ltd, 130 Milton Park, Abingdon, Oxon OX14 4SB. Telephone: (44) 01235 827720. Fax: (44) 01235 400454. Lines are open from 9.00 – 5.00, Monday to Saturday, with a 24-hour message answering service. Visit our website at www.hoddereducation.co.uk. Hodder Gibson can be contacted direct on: Tel: 0141 848 1609; Fax: 0141 889 6315; email: hoddergibson@hodder.co.uk

First Edition published in 2001
Second Edition first published in 2006 by
Hodder Gibson, an imprint of Hodder Education, an Hachette UK Company
2a Christie Street
Paisley PA1 1NB

ISBN-13: 978-0-340-91211-9

Impression number 10 9 8 7 6 5 4 3 2
Year 2010 2009 2008 2007 2006

ISBN-13: 978-0-340-91212-6 (with Answers)

Impression number 10 9 8 7 6 5 4
Year 2010 2009

Cover photo by TEK Image/Science Photo Library
Typeset in Minion 12pt by Fakenham Photosetting Limited
Printed in Italy.

A catalogue record for this title is available from the British Library

Contents

Preface

'First figure out why you want the students to learn the subject and what you want them to know, and the method will result more or less by chance'
Richard Feynman, Nobel Laureate

This second edition of our book tries to build on the key features of the original edition, namely a clear exposition of the key points of the course while allowing students review clear examples and test themselves on suitable questions. To this edition we have added examples of physics from many situations connected to the topics in the course. The changes to the arrangements are all included and the use of colour is useful in many diagrams and photographs. Many more questions have been added at appropriate points in the chapters.

We are grateful to Rothwell Glen for his unstinting work as editor. He has helped us to clarify many topics and his lucid comments have been welcome. Students and colleagues have raised questions which have altered our own perceptions and we hope that the book reflects these concerns. John Mitchell and Katherine Bennett of Hodder Gibson have helped produce an excellent book.

Finally our families and friends have borne a large burden for the time involved in such a task.

Andrew McCormick
Arthur Baillie
Rothwell Glen

May 2006

Electricity and Electronics

Electricity is used for many everyday things such as cooking, cleaning, transport and entertainment. The types of electrical circuits used and the measurement of current, voltage, resistance, power and energy are discussed. This unit also looks at the practical uses of electromagnetism and a number of common electrical components.

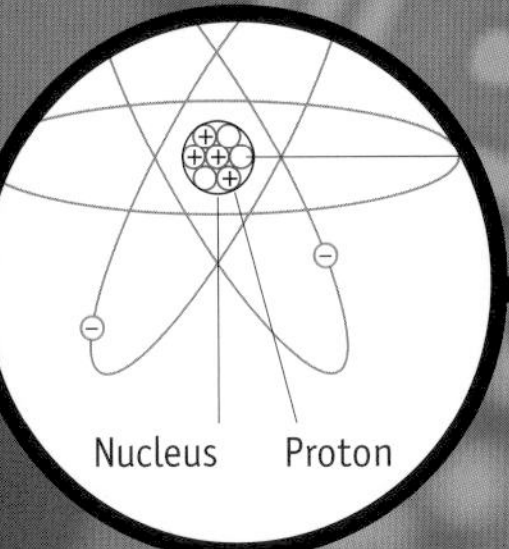

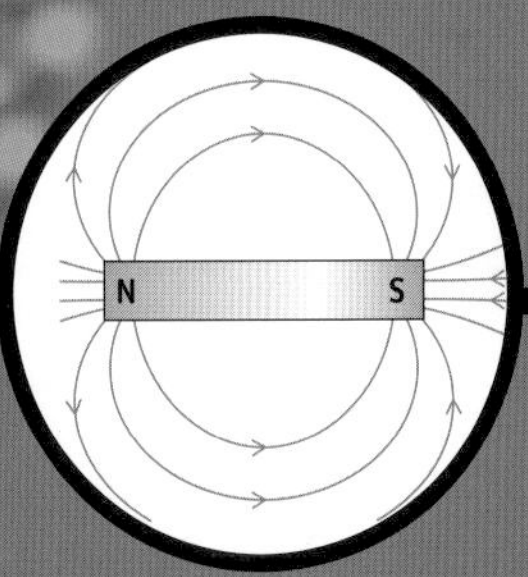

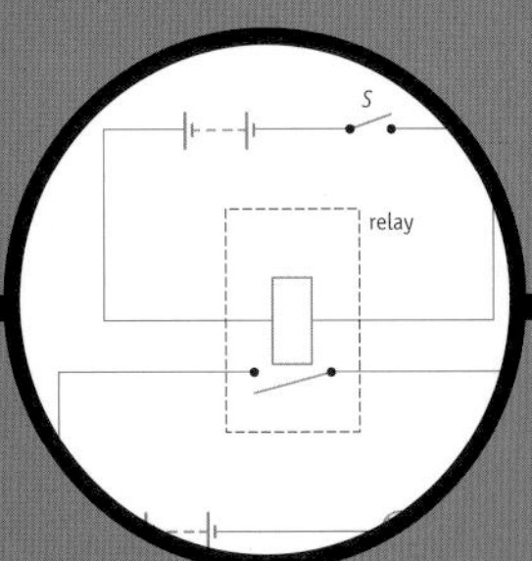

1 Circuits

At the end of this chapter you should be able to...

1. State that electrons are free to move in a conductor.
2. Describe electrical current in terms of the movement of charges around a circuit.
3. Carry out calculations involving the relationship between charge, current and time.
4. Distinguish between conductors and insulators and give examples of each.
5. Draw and identify circuit symbols for an ammeter, voltmeter, battery, resistor, variable resistor, fuse, switch and lamp.
6. State that the voltage of a supply is a measure of the energy given to the charges in a circuit.
7. State that an increase in the resistance of a circuit leads to a decrease in the current in that circuit.
8. Draw circuit diagrams to show the correct positions of an ammeter and voltmeter in a circuit.
9. State that in a series circuit the current is the same at all positions.
10. State that the sum of the potential differences across the components in series is equal to the voltage of the supply.
11. State that the sum of the currents in parallel branches is equal to the current drawn from the supply.
12. State that the potential difference across components in parallel is the same for each component.
13. State that V/I for a resistor remains approximately constant for different currents.
14. Carry out calculations involving the relationship between potential difference, current and resistance.
15. Carry out calculations involving resistors connected in series and parallel.
16. State that a potential divider consists of a number of resistors, or a variable resistor, connected across a supply.
17. Carry out calculations involving potential differences and resistances in a potential divider.

What is electricity?

All solids, liquids and gases are made up of atoms. An atom consists of a positively charged centre or nucleus surrounded by a 'cloud' of rapidly revolving negative charges called electrons. The nucleus is made of particles called protons (positively charged) and neutrons (uncharged).

Charge is measured in coulombs (C). The charge on a proton is 1.6×10^{-19} C. The charge on an electron is the same size as the charge on a proton.

Charge, current and time

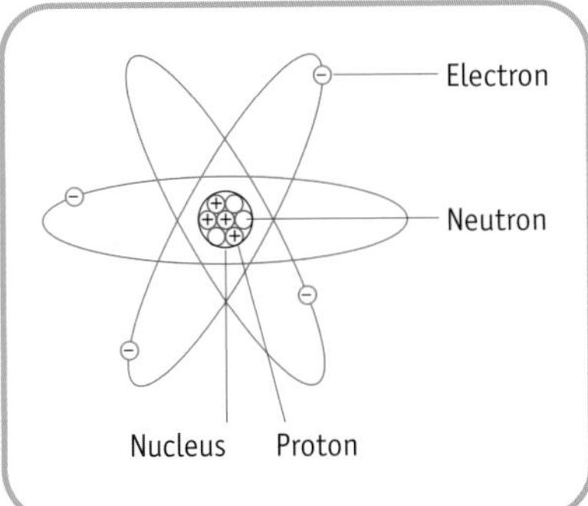

Figure 1.1 *Model of an atom*

Consider a simple electrical circuit – a lamp connected to a battery. The lamp lights up. This is due to electrons from the negative terminal of the battery moving through the wires and lamp to the positive terminal of the battery. This movement of negative charges is called an electric current (or current for short). (see Figure 1.2).

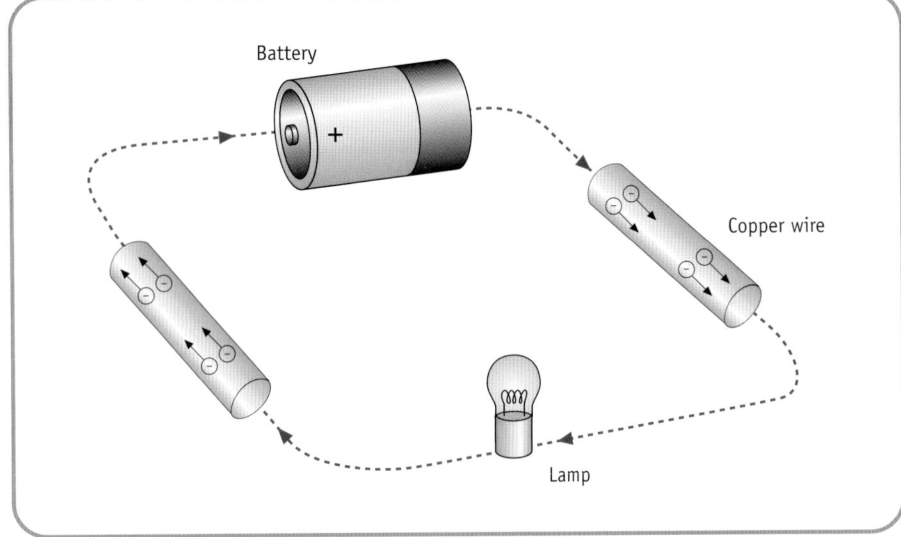

Figure 1.2 *An electric current is the movement of electrons from the negative to the positive terminals of a source of electrical energy such as a battery*

The amount of charge transferred is given by:

$$\text{charge transferred} = \text{current} \times \text{time}$$

$$Q = It$$

where Q = charge transferred, I = current, t = time.

Charge is measured in coulombs (C), current is measured in amperes (A) and time in seconds (s).

1 coulomb = 1 ampere second (1 C = 1 A s)

Example

The current in a wire is 0.2 A. Calculate the charge transferred through the wire in 50 seconds.

Solution

$Q = It$

$Q = 0.2 \times 50$

$Q = 10$ C

Example

In a time of 10 minutes, 2400 coulombs of charge pass through a lamp. Calculate the current in the lamp.

Solution

$Q = It$

$2400 = I \times (10 \times 60)$

$2400 = I \times 600$

$I = \frac{2400}{600} = 4\text{ A}$

Example

When 600 C of charge are transferred through a wire, the current in the wire is 1.5 A. Calculate the time taken for the charge to be transferred?

Solution

$Q = It$

$600 = 1.5 \times t$

$t = \frac{600}{1.5} = 400\text{ s}$

Conductors and insulators

Electrons can only move from the negative terminal to the positive terminal of a battery if there is an electrical path between them. Materials which allow electrons to move through them easily, to form an electric current, are known as conductors. Conductors are mainly metals, such as copper, gold and silver. However, carbon is also a good conductor.

Materials which do not allow electrons to move through them easily are called insulators. Glass, plastic, wood and air are examples of insulators.

Did you know?

During certain weather conditions, clouds can store very large amounts of electrical charge. The charged clouds can discharge to the ground – a lightning strike – usually via tall buildings or trees. A lightning strike can cause considerable damage to a building. Because of this many tall buildings are fitted with a lightning conductor – a lightning conductor consists of a thick copper strip connected to the ground. In the event of a lightning strike the charge is conducted safely to the ground.

Figure 1.3 A lightning conductor

Voltage or potential difference

A battery changes chemical energy into electrical energy. This electrical energy is carried by the electrons that move round the circuit and converted into other forms of energy, e.g. heat and light, by components in the circuit such as a lamp. The amount of electrical energy the electrons have at any point in a circuit is known as their 'potential'. As electrons move between two points in an electric circuit, they transfer electrical energy into other forms of energy. This means that the electrons have a different amount of electrical energy (or potential) at the two points. There is a potential difference (or p.d.) between the points.

The voltage or potential difference between the two terminals of the battery is a measure of the electrical energy given to the electrons by the battery. To be exact, the voltage or p.d. of a battery is the electrical energy given to one coulomb of charge passing through the battery. For example a 6-volt battery gives four times as much electrical energy to each coulomb of charge passing through it compared to a 1.5-volt battery.

Circuit symbols

The circuit symbols for a battery, resistor, variable resistor, fuse, switch, lamp, ammeter and voltmeter are shown in Figure 1.4.

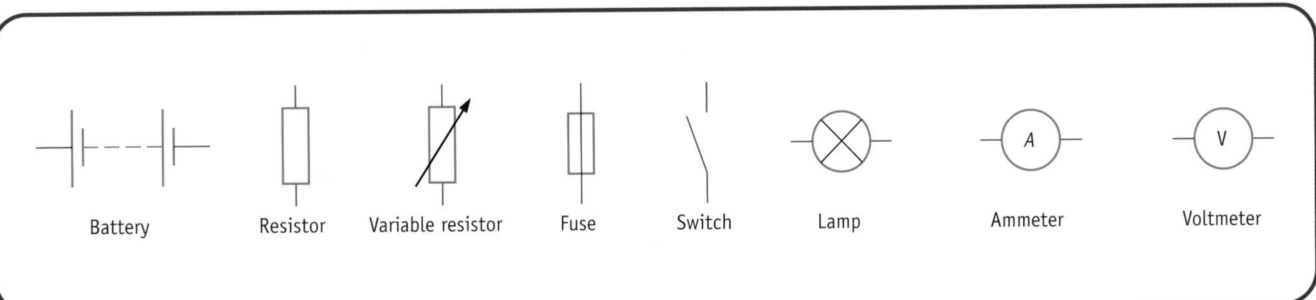

Figure 1.4 *Common circuit symbols*

Measuring current

An ammeter is used to measure electric current. It is measured in amperes (A). Figure 1.5 shows how an ammeter is connected in an electrical circuit.

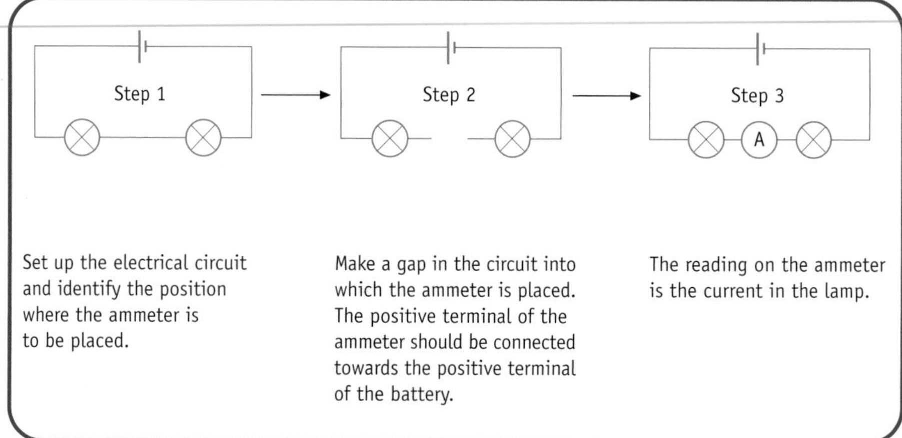

Figure 1.5 *Connecting an ammeter into an electrical circuit*

An ammeter measures the current (i.e. 'counts' the number of coulombs each second) **in** a component.

Measuring voltage

A voltmeter is used to measure voltage or potential difference (p.d.). It is measured in volts (V). Figure 1.6 shows how a voltmeter is connected to an electrical circuit.

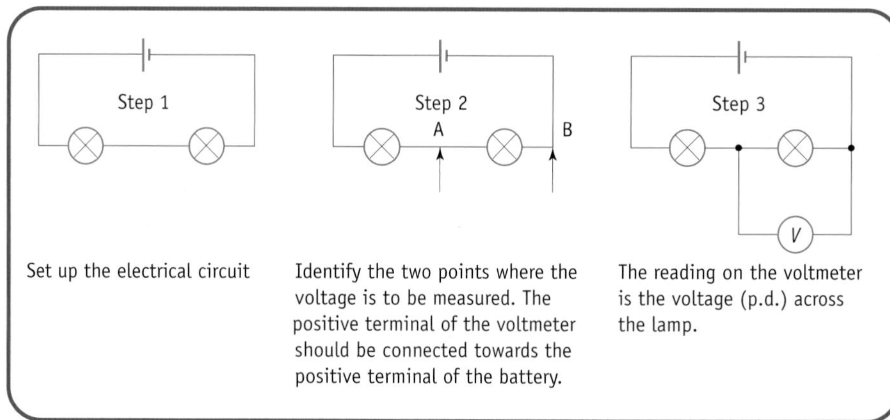

Figure 1.6 *Connecting a voltmeter into an electrical circuit*

A voltmeter measures the voltage or p.d. **across** a component (i.e. the number of joules of energy transferred by each coulomb of charge).

Ammeters and voltmeters can be connected to the same circuit using the instructions given above.

Resistance

Most materials oppose current passing through them. This opposition to the current is called resistance. Resistance is measured in ohms (Ω).

For most materials resistance depends on the:

- Type of material – the better the conductor the lower the resistance.
- Length of the material – the longer the material the higher the resistance.
- Thickness of the material – the thinner the material the higher the resistance.
- Temperature of the material – the higher the temperature the higher the resistance.

For a resistor, the ratio of $\frac{\text{voltage (p.d.) across resistor}}{\text{current in resistor}}$ remains approximately constant.

This constant is the resistance of the resistor. Therefore, the resistance of a resistor can be calculated provided the voltage or potential difference across the resistor and the current in the resistor are known. Then:

$$\text{resistance} = \frac{\text{voltage (p.d.) across resistor}}{\text{current in resistor}} \quad \text{i.e. } R = \frac{V}{I}$$

This is known as Ohm's law.

$$\text{In units: ohms } (\Omega) = \frac{\text{volts (V)}}{\text{amperes (A)}}$$

Ohm's law is normally written as

Voltage (p.d.) across resistor = current in resistor x resistance of resistor

Or $V = IR$

The resistance of the lamp shown in Figure 1.7 can be calcualted using the voltmeter and ammeter readings:

$$\text{Resistance of lamp} = \frac{\text{voltage across lamp}}{\text{current in lamp}} = \frac{\text{reading on voltmeter}}{\text{reading on ammeter}}$$

Figure 1.7 *The resistance of the resistor (in this case a lamp) can be calculated using the voltmeter and ammeter readings.*

Example

There is a current of 1.5 A in a lamp. The lamp has a resistance of 8 Ω. Calculate the potential difference across the ends of the lamp.

Solution

$V = IR$

$V = 1.5 \times 8$

$V = 12$ V

Did you know?

At room temperature, all materials have resistance. As most materials are cooled to lower and lower temperatures, their resistance decreases. When at very low temperatures of about −140°C the electrical resistance of some metals and alloys becomes zero. These materials are known as superconductors. Superconductors are ideal in applications where you don't want electrical energy changed into heat. This is put to use in very strong electromagnets where an intense magnetic field is required.

Example

The voltage across an electric motor is 6 V. The current in the motor is 0.3 A. Calculate the resistance of the motor.

Solution

$V = IR$

$6 = 0.3 \times R$

$R = \dfrac{6}{0.3} = 20\ \Omega$

Example

A kettle is connected to a 230 V supply and switched on. The element of the kettle has a resistance of 25 Ω. Calculate the current in the element.

Solution

$V = IR$

$230 = I \times 25$

$I = \dfrac{230}{25} = 9.2\ \text{A}$

A resistor whose resistance can be changed is known as a variable resistor. The resistance is changed by altering the length of the wire in the resistor (the longer the wire, the higher the resistance) – the higher the resistance the smaller the current. Variable resistors are often used as volume or brightness controls on televisions, and dimmers on lights.

Types of circuit

Electrical components, such as lamps and resistors, can be connected in series, in parallel or a mixture of series and parallel. A series circuit has only one electrical path from the negative terminal of the battery to the positive terminal. A parallel circuit has more than one electrical path from the negative terminal of the battery to the positive terminal. Figure 1.8 shows three lamps connected in series, while Figure 1.9 shows three lamps connected in parallel. Figure 1.10 shows a mixed series and parallel circuit in which a lamp is connected in series with two resistors which are connected in parallel.

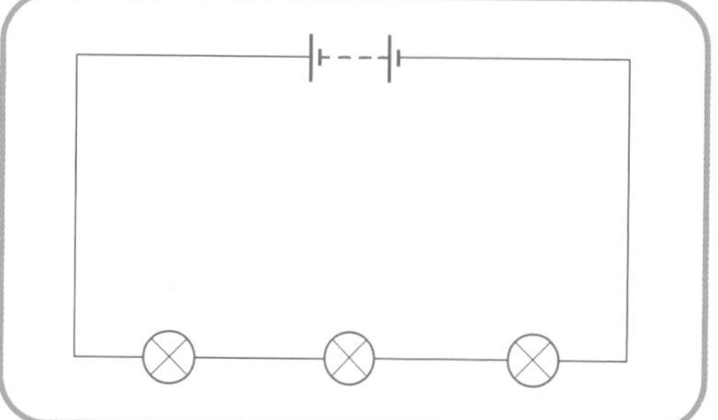

Figure 1.8 *A series circuit*

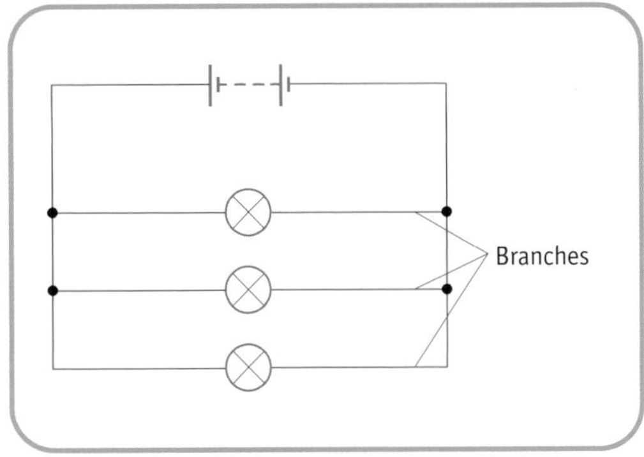

Figure 1.9 *A parallel circuit*

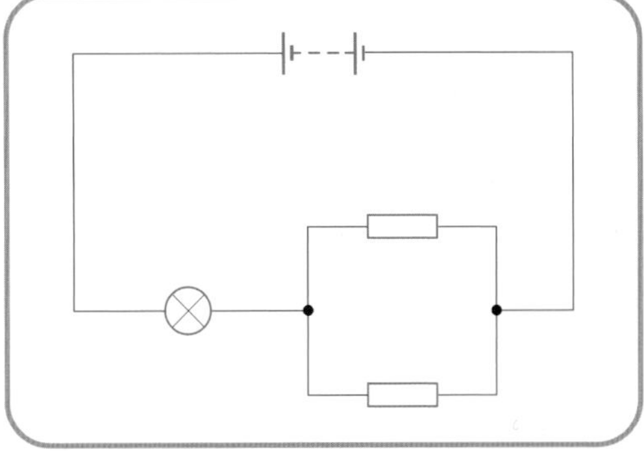

Figure 1.10 *A mixed series and parallel circuit*

A series circuit

Three resistors of value 1 Ω, 4 Ω and 3 Ω are connected in series as shown in Figure 1.11. Ammeters are connected to measure the current at various positions. Voltmeters have also been connected to measure the voltage (p.d.) across each resistor and the battery. The table shows the readings on the meters and the values of the resistances calculated using Ohm's law.

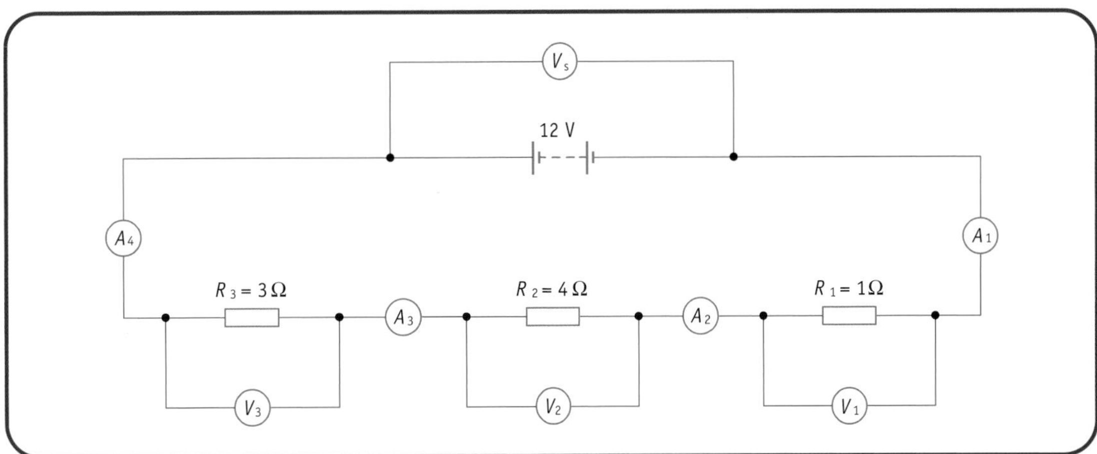

Figure 1.11 *Measuring the current in and the voltage (p.d.) across resistors connected in series*

Reading on ammeter	Reading on voltmeter	Resistance of resistor
A_1 = 1.5 A	V_1 = 1.5 V	R_1 = 1 Ω
A_2 = 1.5 A	V_2 = 6.0 V	R_2 = 4 Ω
A_3 = 1.5 A	V_3 = 4.5 V	R_3 = 3 Ω
A_4 = 1.5 A	–	–
–	V_s = 12 V	–

From the table we can conclude that:

- The current at different positions is the same.
- The sum of the voltages (p.d.s) across the resistors is equal to the supply voltage.

Figure 1.12 shows a single resistor connected to the same supply voltage. This circuit must have the same combined or total resistance as Figure 1.11 since it has the same supply voltage and the same circuit current. Using Ohm's law on this circuit gives a value for the single resistor of 8 Ω (as $R = V_s/I = 12/1.5 = 8\ \Omega$).

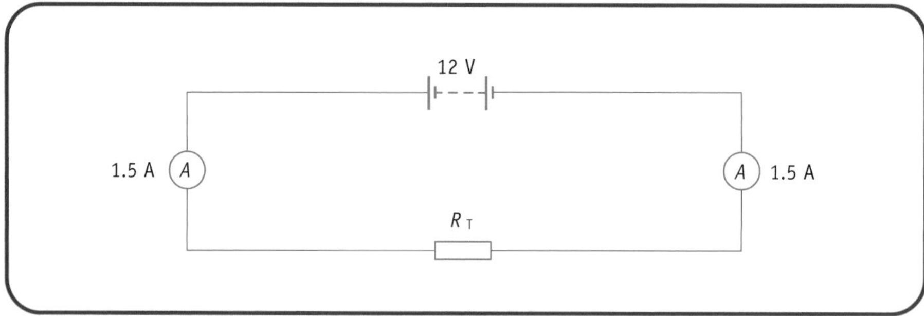

Figure 1.12 *Identical circuit to that shown in Figure 1.11*

Comparing Figures 1.11 and 1.12 we see that the combined or total resistance of a number of resistors connected in series is equal to the sum of the individual resistances. Adding resistors in series increases the total resistance of the circuit.

For the series circuit shown in Figure 1.13 we have:

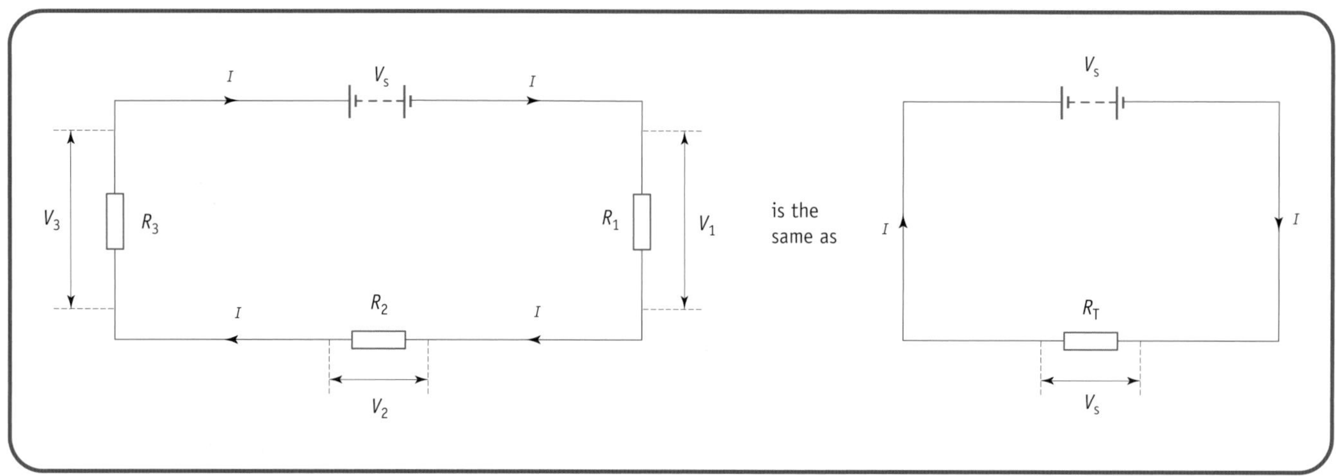

Figure 1.13 *Identical circuits provided* $R_T = R_1 + R_2 + R_3$

- The current is the same at all positions – current does not split up.
- The supply voltage is equal to the sum of the voltages (potential differences) across the components, i.e. $V_s = V_1 + V_2 + V_3$.
- The total resistance (R_T) of the circuit is found using $R_T = R_1 + R_2 + R_3$.

Note. For a series circuit the total resistance is greater than the value of the largest resistance connected in series.

Example

Three resistors of value 100 Ω, 47 Ω and 33 Ω are connected in series. What is the total resistance of the resistors?

Solution

$R_T = R_1 + R_2 + R_3$

$R_T = 100 + 47 + 33$

$R_T = 180\ \Omega$

Example

For the circuit shown in Figure 1.14, calculate the readings on the ammeters A_1, A_2 and the potential difference across each of the resistors.

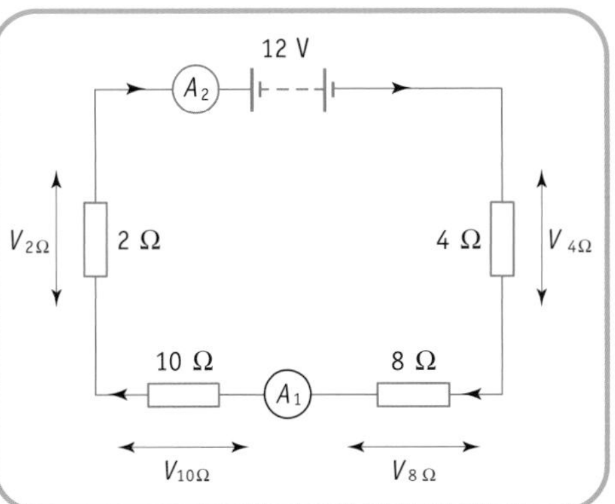

Figure 1.14 *Example of a series circuit*

Solution

Total resistance of circuit, $R_T = R_1 + R_2 + R_3 + R_4$

$R_T = 4 + 8 + 10 + 2$

$R_T = 24\ \Omega$

From Ohm's law: Circuit current $= I = \dfrac{Vs}{R_T} = \dfrac{12}{24} = 0.5$ A

$A_1 = A_2 = 0.5$ A (since current in series circuit is the same at all points)

The potential differences across each resistor are calculated using Ohm's law:

Voltage (p.d.) across resistor = current in resistor x resistance of resistor

$V_{4\Omega} = I \times R_{4\Omega} = 0.5 \times 4 = 2$ V

$V_{8\Omega} = I \times R_{8\Omega} = 0.5 \times 8 = 4$ V

$V_{10\Omega} = I \times R_{10\Omega} = 0.5 \times 10 = 5$ V

$V_{2\Omega} = I \times R_{2\Omega} = 0.5 \times 2 = 1$ V

A parallel circuit

Two resistors of value 6 Ω and 12 Ω are connected in parallel as shown in Figure 1.15. Ammeters are connected to measure the current at various positions. Voltmeters have also been connected to measure the potential difference across each resistor and the battery. The table shows the readings on the meters and the values of the resistances calculated using Ohm's law.

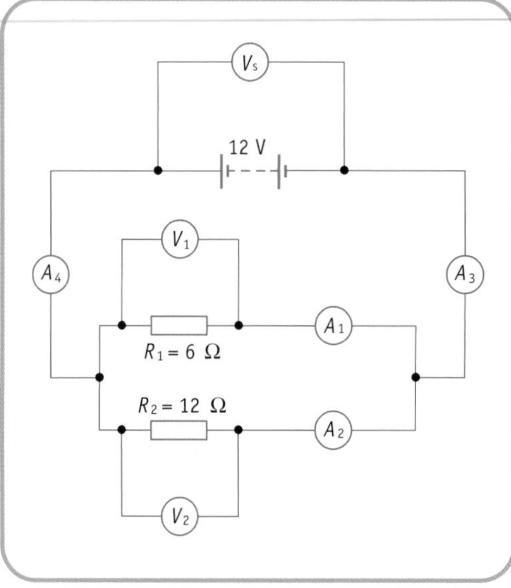

Figure 1.15 *Measuring the current in and the voltage (p.d.) across resistors connected in parallel*

Reading on ammeter	Reading on voltmeter	Resistance of resistor
$A_1 = 2$ A	$V_1 = 12$ V	$R_1 = 6\ \Omega$
$A_2 = 1$ A	$V_2 = 12$ V	$R_2 = 12\ \Omega$
$A_3 = 3$ A	–	–
$A_4 = 3$ A	–	–
–	$V_s = 12$ V	–

From the table we can conclude that:

- The current from the supply is equal to the sum of the currents in the branches.
- The voltages (p.ds) across the resistors connected in parallel are the same.

Figure 1.16 shows a single resistor connected to the same supply voltage. This circuit must have the same combined or total resistance as Figure 1.15 since it has the same supply voltage and the same current from the supply. Using Ohm's law on this circuit gives a value for the single resistor of 4 Ω (as $R = V_s/I = 12/3 = 4\ \Omega$).

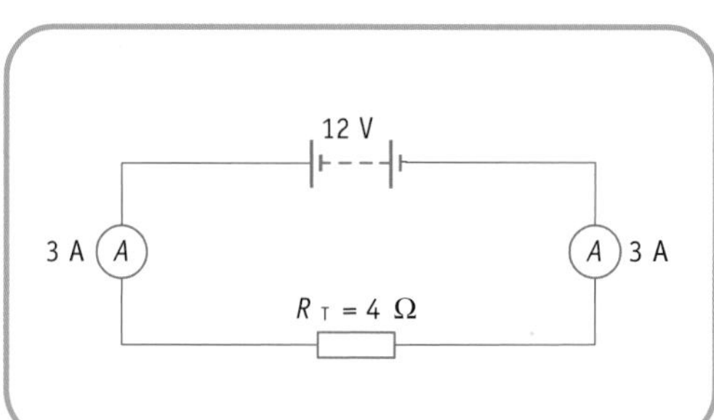

Figure 1.16 *Identical circuit to that shown in Figure 1.15*

Comparing Figures 1.15 and 1.16 we see that the combined resistance of a number of resistors connected in parallel is smaller than any of the individual resistances.

The combined resistance of 4 Ω is obtained for these resistors as follows:

$$\frac{1}{R_T} = \frac{1}{R_1} + \frac{1}{R_2} = \frac{1}{6} + \frac{1}{12} = 0.167 + 0.083 = 0.25$$

$$\frac{1}{R_T} = 0.25$$

$$R_T = \frac{1}{0.25} = 4\ \Omega$$

For the parallel circuit shown in Figure 1.17 we have:

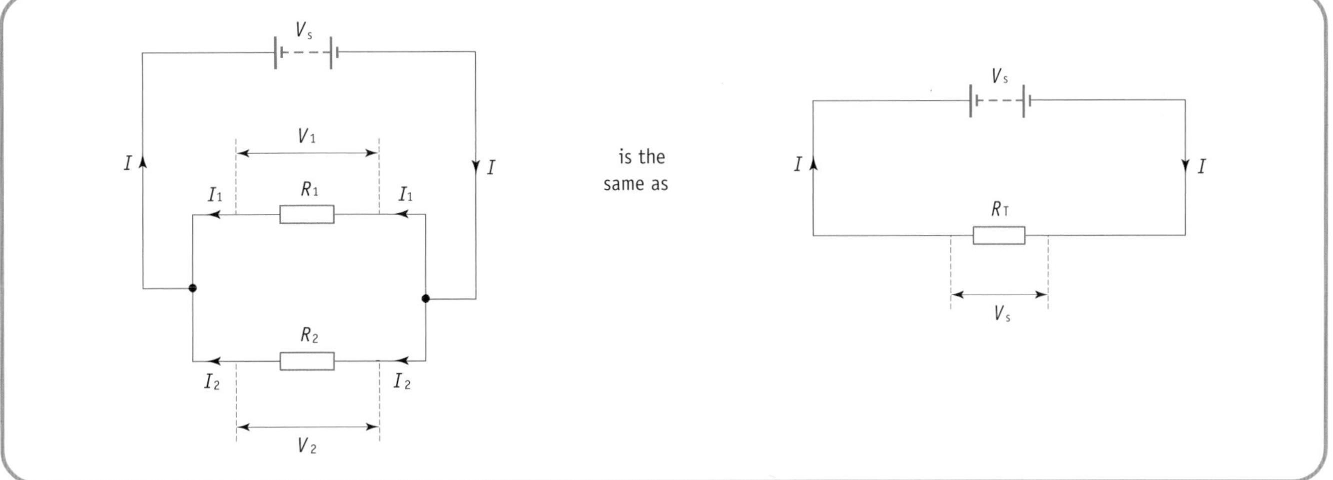

Figure 1.17 *Identical circuits provided –* $\frac{1}{R_T} = \frac{1}{R_1} + \frac{1}{R_2}$

- Current from supply = sum of currents in the branches i.e. $I = I_1 + I_2$.
- The voltages (p.ds) across resistors are the same, i.e. $V_1 = V_2$ and in this case $V_s = V_1 = V_2$.
- The total resistance of the circuit is found using $\frac{1}{R_T} = \frac{1}{R_1} + \frac{1}{R_2}$

Note. For a parallel circuit, the total resistance is less than the value of the smallest resistance connected in parallel.

Example

Three resistors of resistance 20 Ω, 60 Ω and 30 Ω are connected in parallel. Calculate the total resistance of the resistors.

Solution

$$\frac{1}{R_T} = \frac{1}{R_1} + \frac{1}{R_2} + \frac{1}{R_3} = \frac{1}{20} + \frac{1}{60} + \frac{1}{30} = 0.05 + 0.017 + 0.033 = 0.1$$

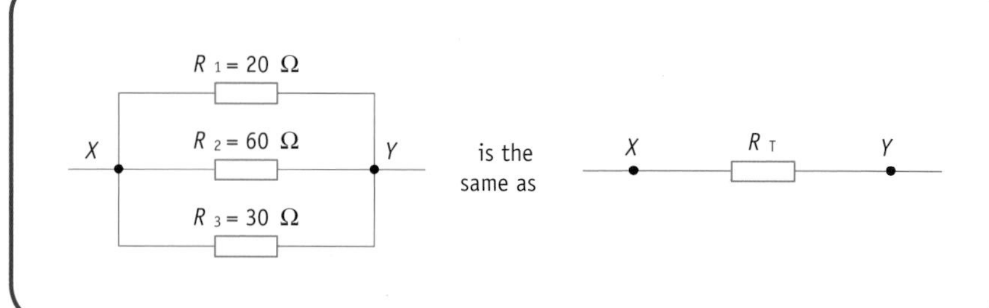

Figure 1.18 *Example of a parallel circuit*

$$\frac{1}{R_T} = 0.1$$

$$R_T = \frac{1}{0.1} = 10\ \Omega$$

As a check we would expect our answer to be smaller than 20 Ω.

Example

Two resistors each of resistance 1 kΩ are connected in parallel. Calculate their total resistance.

Solution

$$\frac{1}{R_T} = \frac{1}{R_1} + \frac{1}{R_2} = \frac{1}{1000} + \frac{1}{1000} = 0.001 + 0.001$$

$$\frac{1}{R_T} = 0.002$$

$$R_T = \frac{1}{0.002} = 500\ \Omega$$

Note that the total resistance of two identical resistors connected in parallel is half that of one of the resistors.

Example

In the circuit shown in Figure 1.19, calculate (a) the total resistance of the circuit, (b) the current drawn from the battery, (c) the p.d. across the 10 Ω resistor.

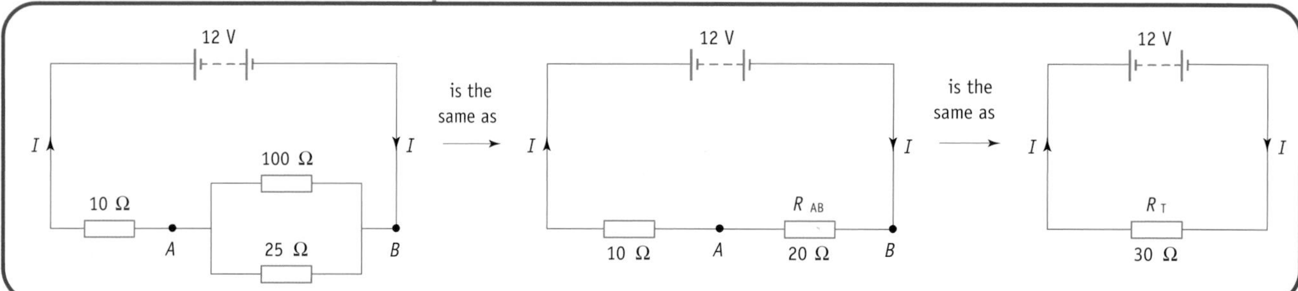

Figure 1.19 *Example of a mixed series and parallel circuit*

Solution

a) $\dfrac{1}{R_{AB}} = \dfrac{1}{R_1} + \dfrac{1}{R_2} = \dfrac{1}{100} + \dfrac{1}{25} = 0.01 + 0.04 = 0.05$

$R_{AB} = \dfrac{1}{0.05} = 20\ \Omega$

Total resistance $R_T = 10 + R_{AB} = 10 + 20 = 30\ \Omega$

b) $V_S = I\ R_T$ gives $12 = I \times 30$

Hence $I = \dfrac{12}{30} = 0.4$ A

c) Voltage (p.d.) across 10 Ω resistor = current in resistor × resistance of resistor

$V = I\ R$

$V = 0.4 \times 10$

$V = 4$ V

Potential dividers

For a series circuit, the supply voltage is equal to the sum of the potential differences (voltages) across the individual resistors, i.e. $V_s = V_1 + V_2$. This means that the supply voltage is split up into, in this case, two smaller bits, and so the supply voltage is divided between the components.

A potential divider consists of two devices, usually resistors, connected in series as shown in Figure 1.20. The p.d. of the supply is divided up into two smaller p.ds.

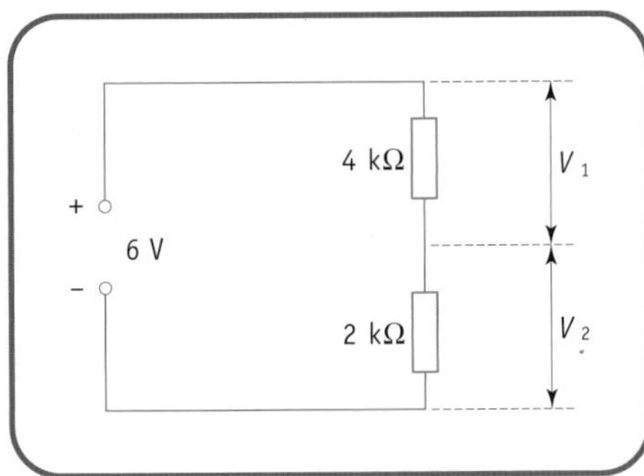

Figure 1.20 *A potential divider circuit*

$R_T = R_1 + R_2$

$R_T = 2000 + 4000 = 6000\ \Omega$

$I_{circuit} = \dfrac{V_S}{R_T} = \dfrac{6}{6000} = 0.001$ A

$V_1 = I_{circuit} \times R_1 = 0.001 \times 2000 = 2$ V

$V_2 = I_{circuit} \times R_2 = 0.001 \times 4000 = 4$ V

Note. The values of V_1 and V_2 depend on the values of R_1 and R_2.

Example

A student sets up an electrical circuit as shown in Figure 1.21.

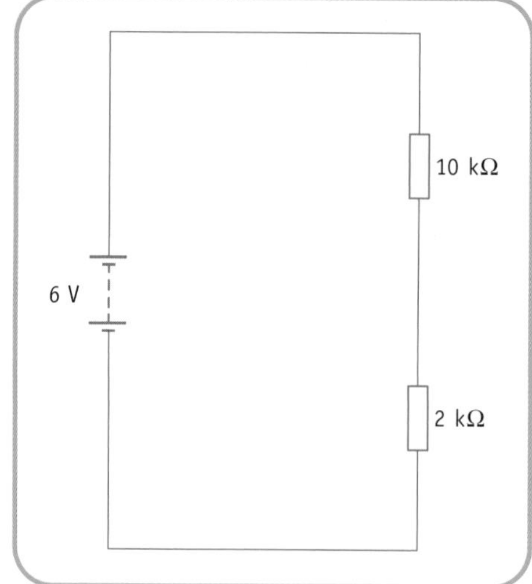

Figure 1.21 *Example of a potential divider*

Calculate the potential difference across the 2 kΩ resistor.

Solution

$R_T = R_1 + R_2$

$R_T = 10\,000 + 2000 = 12\,000\ \Omega$

$$I_{circuit} = \frac{V_s}{R_T} = \frac{6}{12\,000} = 5 \times 10^{-4}\ \text{A}$$

$$V_{2\ k\Omega} = I_{circuit} \times R_{2\ k\Omega} = 5 \times 10^{-4} \times 2000 = 1.0\ \text{V}$$

Note in tackling potential divider problems you usually have to find the circuit current. This can be found from either $I_{circuit} = \frac{V_{supply}}{R_T}$

or $I_{circuit} = I_{component} = \frac{V_{component}}{R_{component}}$

Which equation you use will depend on the information given in the question.

Physics facts and key equations for circuits

- Charge is measured in coulombs (C), current in amperes (A) and time in seconds (s).
- Charge transferred = current x time, i.e. $Q = It$.
- The voltage of a supply is a measure of the energy given to one coulomb of charge.
- An ammeter is connected in series with the component.
- A voltmeter is connected in parallel across the component.

- Voltage (or p.d.) is measured in volts (V), current in amperes (A) and resistance in ohms (Ω).
- Voltage (p.d.) across a resistor = current in resistor × resistance of resistor, i.e. $V = IR$ – this is known as Ohm's law.
- The resistance of a resistor remains constant for different currents provided the temperature of the resistor does not change
- In a series circuit:
  - the current is the same at all points, i.e. $I_1 = I_2 = I_3$
  - the supply voltage is equal to the sum of the voltages (p.d.'s) across components, i.e. $V_s = V_1 + V_2 + V_3$
  - the total (or combined) resistance is found using $R_T = R_1 + R_2 + R_3$
  - the total resistance is greater than the value of the largest resistance conneted in series.

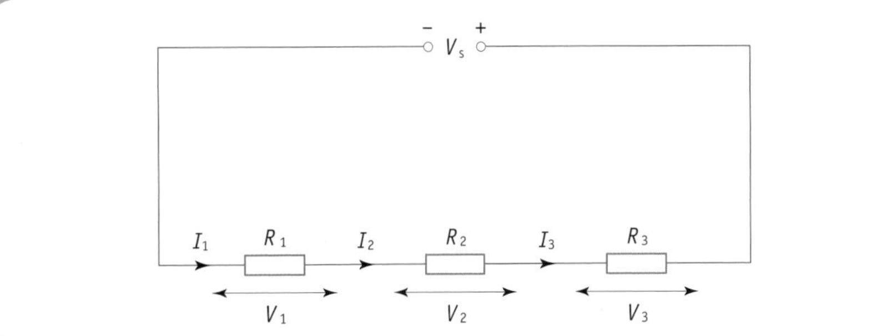

Figure 1.22 *A series circuit*

- In a parallel circuit:
  - the current from the supply is equal to the sum of the currents in the branches, i.e. $I_C = I_1 + I_2 + I_3$
  - the voltage (p.d.) across components is the same, i.e. $V_S = V_1 = V_2 = V_3$
  - the total (or combined) resistance is found by $\frac{1}{R_T} = \frac{1}{R_1} + \frac{1}{R_2} + \frac{1}{R_3}$
  - the total resistance is less than the value of the smallest resistance connected in parallel.

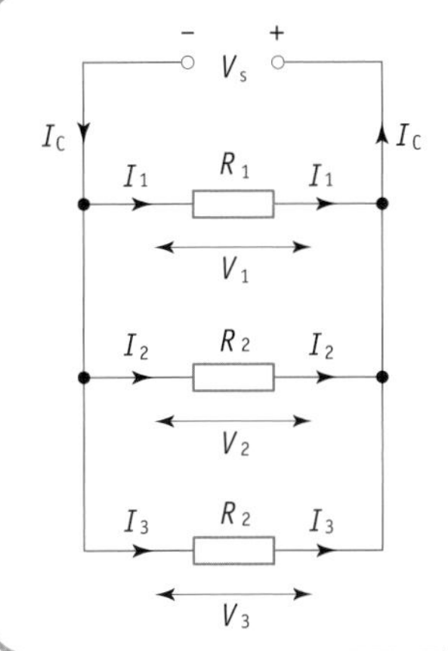

Figure 1.23 *A parallel circuit*

Questions

1. When a kettle is switched on, there is a current of 10 A in the element. The kettle is switched on for 2 minutes. How much charge flows through the element in this time?
2. Calculate the time taken for a current of 2.0 A to transfer 500 C of charge through a lamp.
3. In 5 minutes, 150 C of charge pass through a resistor. Calculate the current in the resistor?

4 An electronic game works from a 9 V supply. What is meant by a supply voltage of 9 V?

5 Draw the circuit symbol for (a) a battery, (b) a lamp, (c) a switch, (d) a resistor, (e) a variable resistor, (f) a fuse, (g) an ammeter and (h) a voltmeter

6 In the circuits shown in Figure 1.24, what are the readings on (a) ammeters A_1, A_2 and A_3, (b) voltmeters V_1, V_2, and V_3?

7 Three resistors of value 47 Ω, 100 Ω and 150 Ω are connected in series. Find the total resistance of these three resistors.

8 Three resistors of value 20 Ω, 20 Ω and 10 Ω are connected in parallel. Find the total resistance of these three resistors.

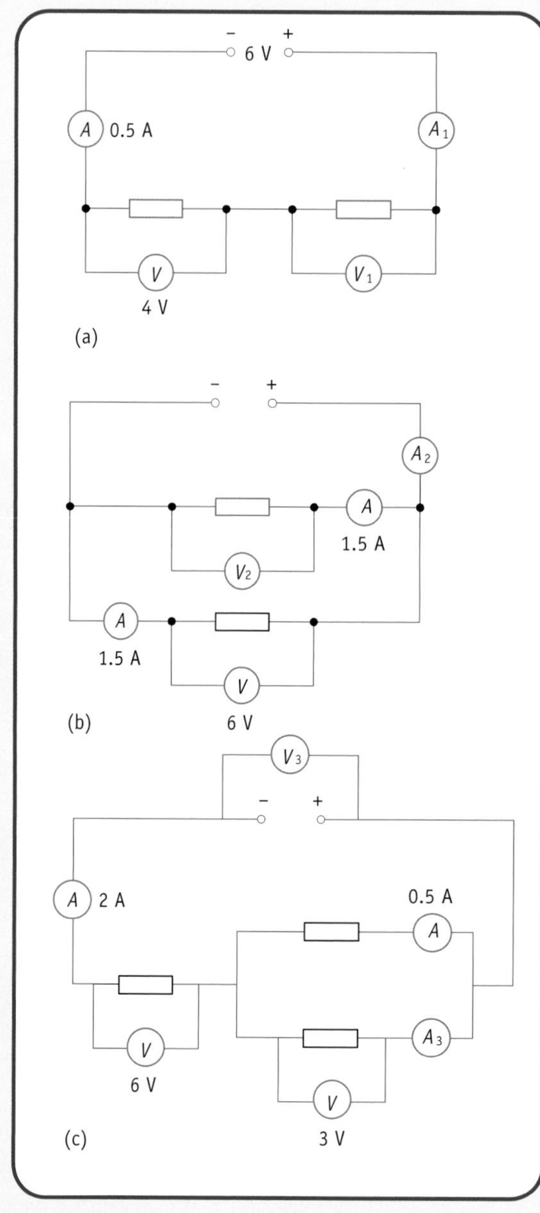

Figure 1.24

9 Four resistors are arranged as shown in Figure 1.25. Calculate the resistance between X and Y.

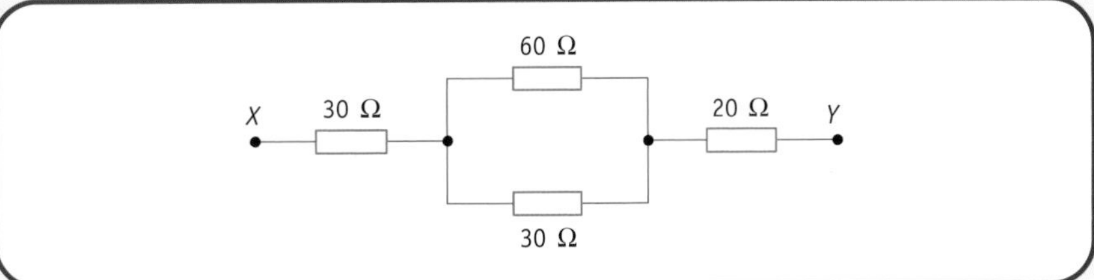

Figure 1.25

10 Redraw each of the diagrams shown in Figure 1.26 to show how **both** a voltmeter is connected to measure the voltage across component *R* **and** an ammeter is connected to measure the current in component *S*.

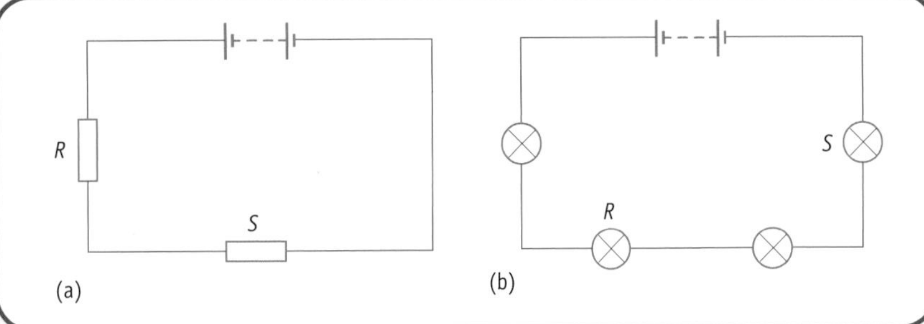

Figure 1.26

11 There is a current of 0.012 A in a 1 kΩ resistor. Calculate the potential difference across the resistor.

12 An electric toaster is connected to the 230 V mains and switched on. There is a current of 4.5 A in the toaster element. Calculate the resistance of the element.

13 The motor of a toy car is connected to a 4.5 V battery. The motor of the car has a resistance of 18 Ω. Calculate the current in the motor.

14 Two resistors, of value 1 kΩ and 3 kΩ, are connected in series with a 9 V battery. What is the potential difference across the 3 kΩ resistor?

2 Electrical energy

At the end of this chapter you should be able to...

1 State that when there is an electrical current in a component, there is an energy transformation.
2 State that the electrical energy transformed each second = VI.
3 Carry out calculations involving the relationships between power, energy, time, current and potential difference.
4 Explain the equivalence between VI, I^2R and V^2/R.
5 Carry out calculations involving the relationships between power, current, voltage and resistance.
6 State that in a lamp electrical energy is transformed into heat and light.
7 State that the energy transformation in an electrical heater occurs in the resistance wire.
8 Explain in terms of current the expressions d.c. and a.c.
9 State that the frequency of the mains supply is 50 Hz.
10 State that the quoted value of an alternating voltage is less than its peak value.
11 State that a d.c. supply and an a.c. supply of the same quoted value will supply the same power to a given resistor.

Power

When there is an electric current in a wire, the electrons making up the current collide with the atoms of the wire. These collisions make the atoms vibrate more and this results in the wire becoming hotter, i.e. electrical energy has been changed into heat in the wire. The amount of heat produced depends on the value of the current and the value of the resistance of the wire. Heating elements for electric fires and kettles change electrical energy into heat in the resistance wire making up the element.

A lamp transfers electrical energy into heat and light in a resistance wire called the filament. The energy transferred in one second is known as the power rating of the lamp.

Power is the energy transferred in one second.

$$\text{Power} = \frac{\text{energy transferred}}{\text{time taken}}$$

$$P = \frac{E}{t}$$

1 watt = 1 joule per second (1 W = 1 J/s)

Example

A hairdryer uses 86.4 kJ of energy in a time of 1.5 minutes. What is the power rating of the hairdryer?

Solution

$$P = \frac{E}{t} = \frac{86.4 \times 10^3}{(1.5 \times 60)} = 960 \text{ W}$$

Power, current and voltage

Three different lamps of known power ratings were connected to an electrical supply. The current in and the voltage (p.d.) across the lamps were recorded. The readings are shown in the table below.

Power rating of lamp (W)	Current in lamp (A)	Voltage across lamp (V)
24	2	12
36	3	12
48	4	12

From the table we can conclude:

Power = current × voltage

$P = IV$

but from Ohm's law $V = IR$

$P = I(IR)$

$P = I^2R$

Alternatively $I = \frac{V}{R}$

$$P = \left(\frac{V}{R}\right)V$$

$$P = \frac{V^2}{R}$$

The equations $P = IV$, $P = I^2R$ and $P = \frac{V^2}{R}$ can be used to find the power rating of appliances.

Use Ohm's Law to calculate the resistance of each of the lamps in the table and then check that the above equations can be used to calculate the power rating of the lamps.

Example

The element of an electric heater is plugged into the 230 V mains supply. The heater is switched on and there is a current of 4.6 A in the element. Calculate the power rating of the element.

Solution

$P = IV = 4.6 \times 230 = 1058$ W

Example

The element of an electric kettle, when operating, has a resistance of 23 Ω. The current in the element is 10 A. Calculate the power rating of the element.

Solution

$P = I^2R = 10^2 \times 23 = 2300$ W

Example

The element of an electric toaster has a power rating of 1050 W. Calculate the resistance of the element when it is operating from the 230 V mains supply.

Solution

$$P = \frac{V^2}{R}$$

$$1050 = \frac{230^2}{R}$$

$$R = \frac{230^2}{1050} = 50.4\ \Omega$$

Direct and alternating current

Oscilloscope

Battery

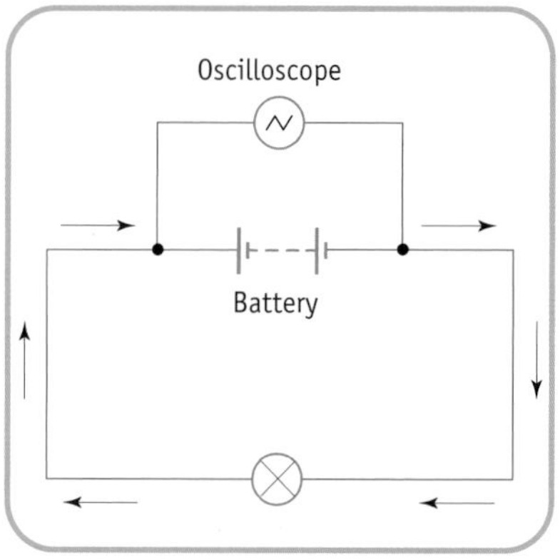

Figure 2.1 *Electrons move in only one direction. This is known as direct current (d.c.)*

Figure 2.1 shows a battery connected to a lamp. Figure 2.2 shows a mains-operated low-voltage power supply connected to an identical lamp. The lamps are equally bright.

In Figure 2.1 electrons (negative charges) move from the negative terminal through the lamp and wires to the positive terminal of the battery. This means that the electrons move in only one direction – this is known as direct current or d.c.

In Figure 2.2 electrons move in one direction, then in the other direction and back again, i.e. the electrons move to and fro. This alternating movement of the electrons is known as alternating current or a.c. The to and fro movement of the electrons is very frequent and occurs 50 times every second and so the frequency of mains electricity is 50 hertz (50 Hz).

The oscilloscope traces from Figures 2.1 and 2.2 are shown in Figures 2.3 and 2.4. The d.c. trace has a constant value of 1.5 V while the a.c. trace alternates from a maximum or peak value of +2.1 V to −2.1 V.

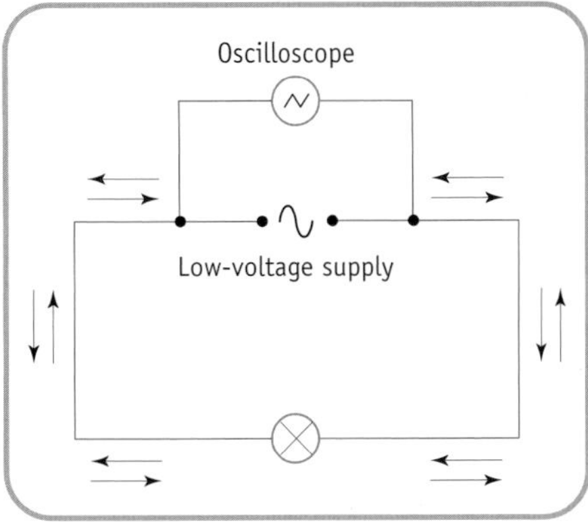

Figure 2.2 *Electron movement is to and fro. This is known as alternating current (a.c.)*

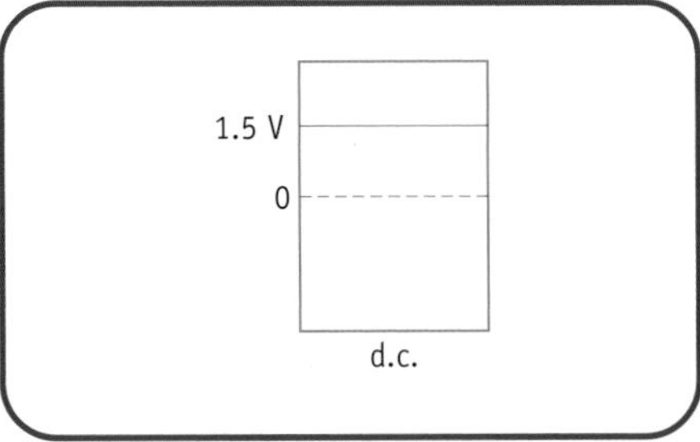

Figure 2.3 *Oscilloscope trace from Figure 2.1*

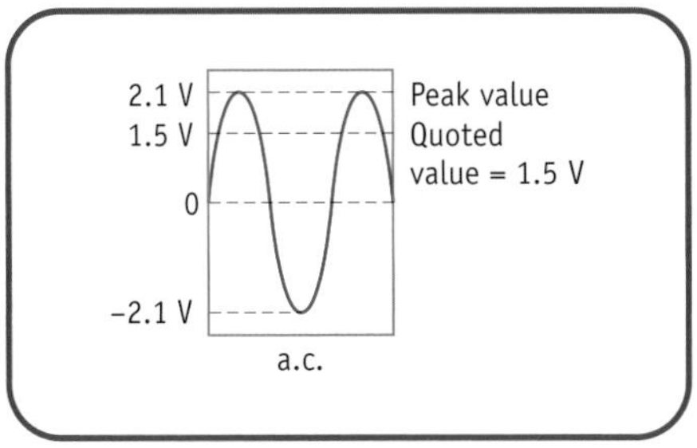

Figure 2.4 *Oscilloscope trace from Figure 2.2*

Since both lamps are equally bright, they must be receiving and giving out the same amount of energy every second, i.e. they have the same power. The a.c. supply of 2.1 V peak is having the same effect as the d.c. supply of 1.5 V. This means that the a.c. supply has an effective or quoted value of 1.5 V.

The quoted (effective) a.c. value is always smaller than the peak a.c. voltage.

The quoted value of the alternating voltage for mains electricity in the UK is 230 volts. However, its peak value is approximately 324 V.

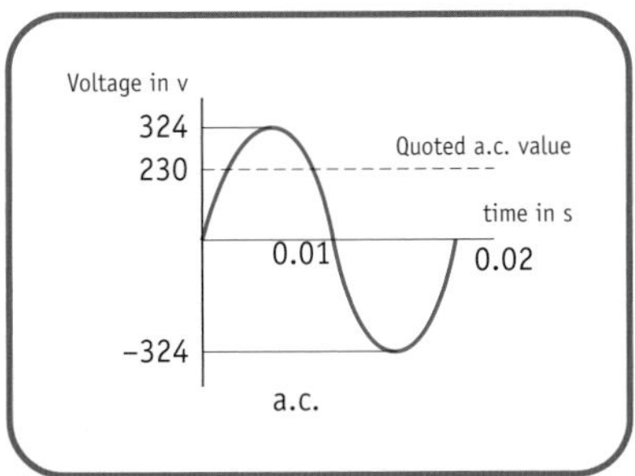

Figure 2.5 *Oscilloscope trace for mains electricity*

Physics facts and key equations for electrical energy

- Power $= \dfrac{\text{energy}}{\text{time}}$, i.e. $P = \dfrac{E}{t}$.
- Power = current × voltage, i.e. $P = IV$.
- Power = current^2 × resistance, i.e. $P = I^2R$.
- Power $= \dfrac{\text{voltage}^2}{\text{resistance}}$, i.e. $P = \dfrac{V^2}{R}$.
- Mains supply has a frequency of 50 Hz.
- The quoted value of mains voltage is 230 V.
- The quoted value of an alternating voltage is less than its peak value.

Questions

1 The element of an electric iron is connected to a 230 V mains supply. The current in the element is 5.5 A.
   a) Calculate the power rating of the element.
   b) How much electrical energy does the element use in one second?

2 A lamp is plugged into a 12 V supply and switched on. When lit the lamp has a resistance of 3 Ω.
   a) Write down the energy change which takes place in the lamp when lit.
   b) Calculate the current in the lamp.
   c) Find the power rating of the lamp.

3 An electric oven is operating from a 230 V mains supply. The element of the oven has a resistance of 17.6 Ω. Calculate the power rating of the element.

4 A spotlight is rated at 12 V, 50 W. The spotlight is switched on and is operating at its rated values. Calculate the resistance of the spotlight.

5 There is a current of 0.6 A in an electric motor. The motor has a power rating of 138 W. Calculate the resistance of the motor when it is operating.

6 An alternating supply has a quoted value of 230 V. Give a possible value for the peak value for this supply.

7 What is the frequency of the mains supply in Great Britain?

3 Electromagnetism

At the end of this chapter you should be able to...

1. State that a magnetic field exists around a current-carrying wire.
2. Identify circumstances in which a voltage will be induced in a conductor.
3. State the factors which affect the size of the induced voltage, i.e. field strength, number of turns on a coil, relative movement.
4. State that transformers are used to change the magnitude of an alternating voltage.
5. Carry out calculations involving input and output voltages, turns ratio and primary and secondary currents for an ideal transformer.

Magnetic fields and electromagnetism

Permanent magnets

A magnet is able to exert a force on certain materials. The region surrounding the magnet is called a magnetic force field or simply a magnetic field. A permanent magnet, as its name implies, has a magnetic field surrounding it which cannot be switched off. The opposite ends, or poles, of a magnet are called north and south (a north pole means a north-seeking pole, i.e. it always wants to point north). The shape of the magnetic field surrounding a magnet can be shown by scattering iron filings on a piece of paper placed on top of it. The direction of the magnetic field can be found using a plotting compass.

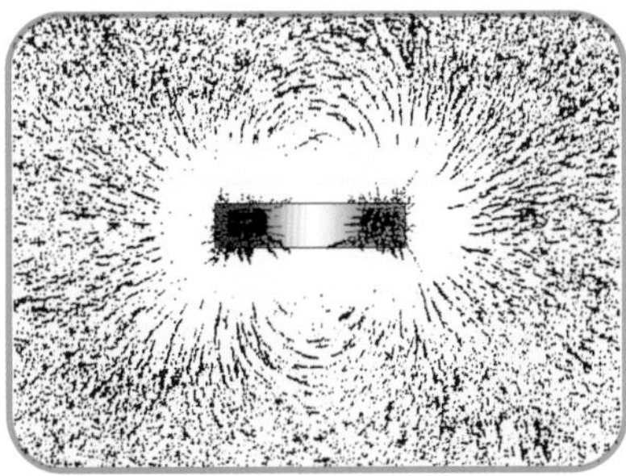

Figure 3.1 *Iron filings show the pattern of the magnetic field lines surrounding a permanent magnet*

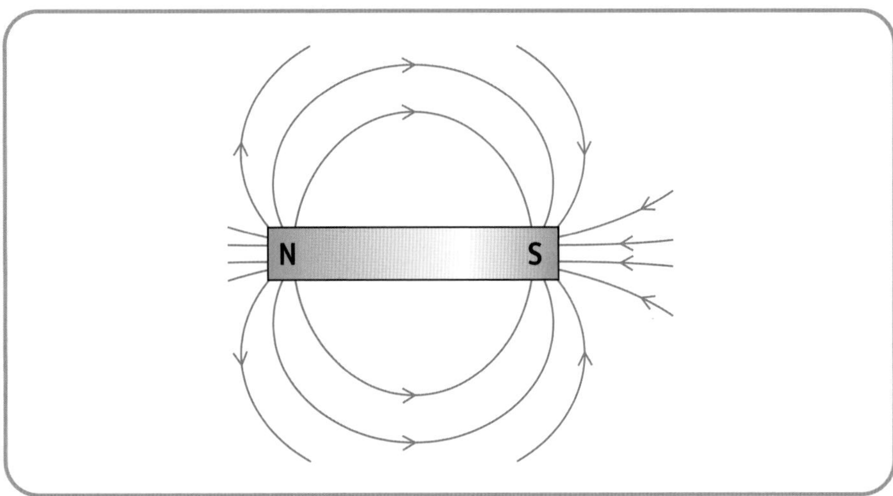

Figure 3.2 *Magnetic field lines surrounding a permanent magnet*

When two permanent magnets are placed close together, their magnetic fields produce forces which cause:

- A north pole to repel a north pole.
- A south pole to repel a south pole.
- A north pole to attract a south pole.

In other words, like poles repel and unlike poles attract.

Some metals such as iron and steel are attracted to a magnet.

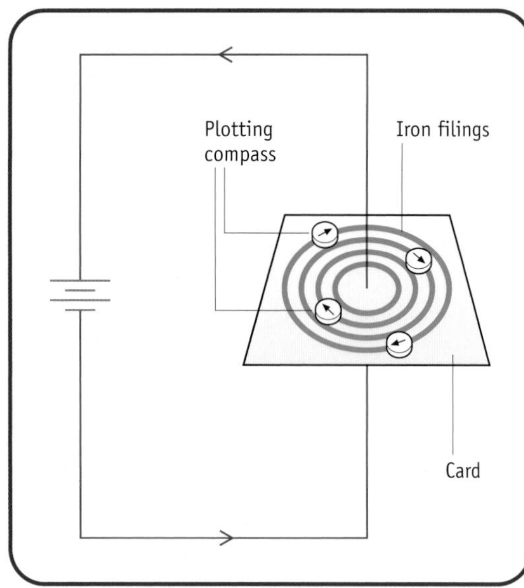

Figure 3.3 *Magnetic field surrounding a current-carrying wire*

Electromagnetism and electromagnets

Figure 3.3 shows a long straight wire passing vertically through a piece of card. A magnetic field surrounds the wire when there is an electrical current in the wire. Increasing the current in the wire increases the strength of magnetic field surrounding the wire. Reversing the direction of the current in the wire reverses the direction of magnetic field around the wire. When there is an electric current in a wire which is coiled around an iron core, the core becomes magnetised and an electromagnet is produced. An electromagnet is shown in Figure 3.4. However, the electromagnet has little strength without the iron core. The iron core is able to concentrate the magnetic field within itself, so giving a stronger magnetic effect.

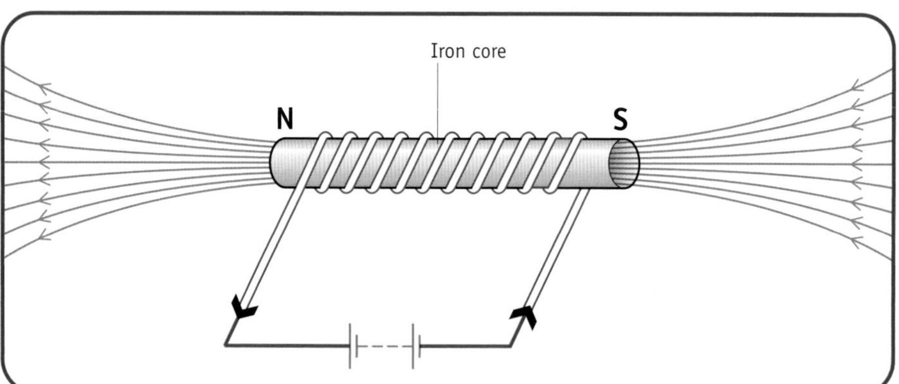

Figure 3.4 *An electromagnet*

The magnetic field of an electromagnet can be made stronger by:

- Increasing the current in the coils of wire.
- Increasing the number of turns of wire on the core.

When an a.c. supply is used, the electromagnet still produces a magnetic field – but one which alternates each time the current changes direction.

When there is no electric current in the coils, there is no magnetic field. This on-off nature of the magnetic field can be used to lift (magnetic field on) and release (magnetic field off) scrap iron as shown in Figure 3.5.

Figure 3.5 *Electromagnets can lift heavy objects*

Did you know?

The strong magnetic field from a superconducting electromagnet is used in some hospitals to detemine what is going on inside the human body. This technique is called Magnetic Resonance Imaging (MRI) and is similar to an X-ray. However, MRI is safer than an X-ray.

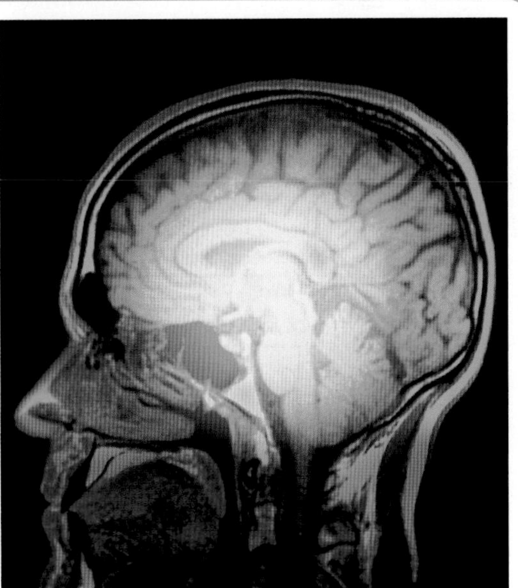

Figure 3.6 *MRI scan of human skull*

Generating electricity

Figure 3.7 shows a coil of wire placed between two magnets. The ends of the coil are connected to a voltmeter. When the coil is moved up or down through the magnetic field, a voltage (p.d.) is produced across the ends of the coil. The size of this voltage (p.d.) is dependent on:

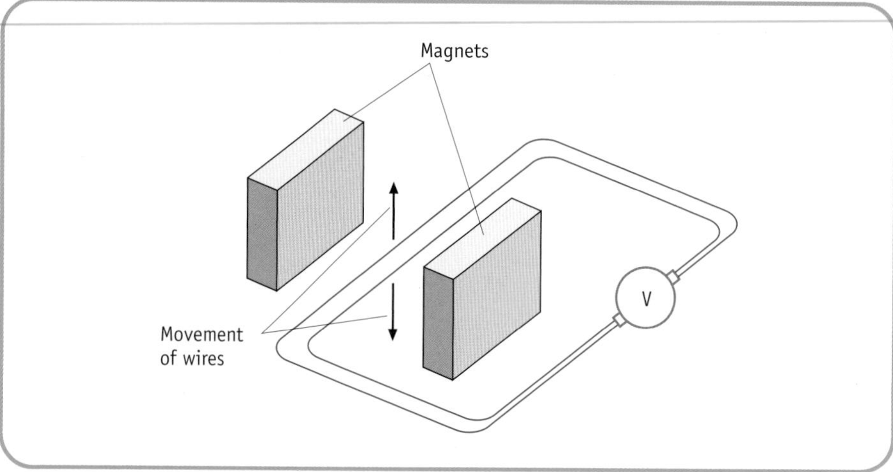

Figure 3.7 *Moving coils of wire through a magnetic field to generate a voltage (p.d.)*

- The number of turns of wire on the coil – the greater the number of turns, the greater the voltage produced.
- The strength of the magnetic field – the stronger the magnetic field, the greater the voltage produced.
- The speed of movement – the faster the coil is moved up or down through the magnetic field, the greater the voltage produced.

Transmitting electrical energy

Figure 3.8 shows how electrical power, generated at a power station, is distributed through the national grid transmission system to our homes.

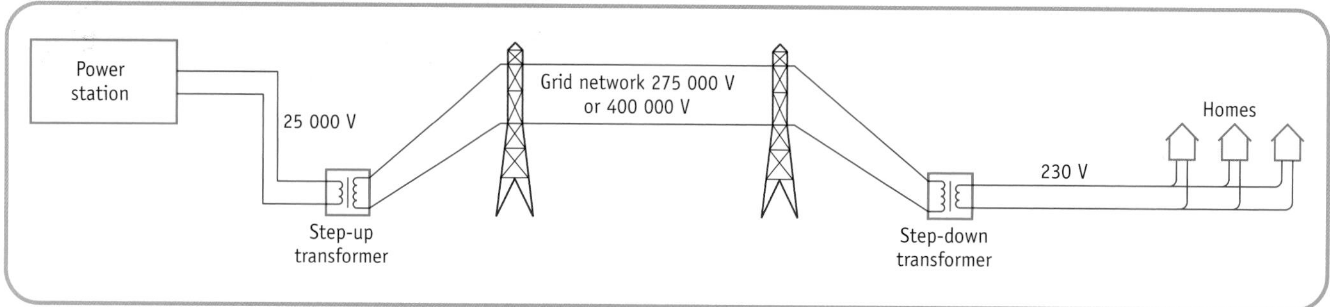

Figure 3.8 *National Grid transmission system*

The generator at the power station produces an output voltage of 25 000 volts. For efficient transmission over long distances this voltage is increased by a (step-up) transformer, to 275 000 volts or 400 000 volts for the national grid system. At the end of the transmission lines a (step-down) transformer reduces the voltage for distribution to consumers.

When electrical power is passed along transmission lines, it is important to keep the power loss as low as possible. Since the power loss from the transmission lines is given by I^2R, the current in the transmission lines, and the resistance of the transmission lines, should both be as low as possible.

In practice there is a limit to how small the resistance of the transmission lines can be. However, transformers make it possible to reduce the size of the current in the transmission lines by increasing the voltage (p.d.). Transformers are therefore essential when electrical energy is to be transmitted over large distances (Figure 3.9).

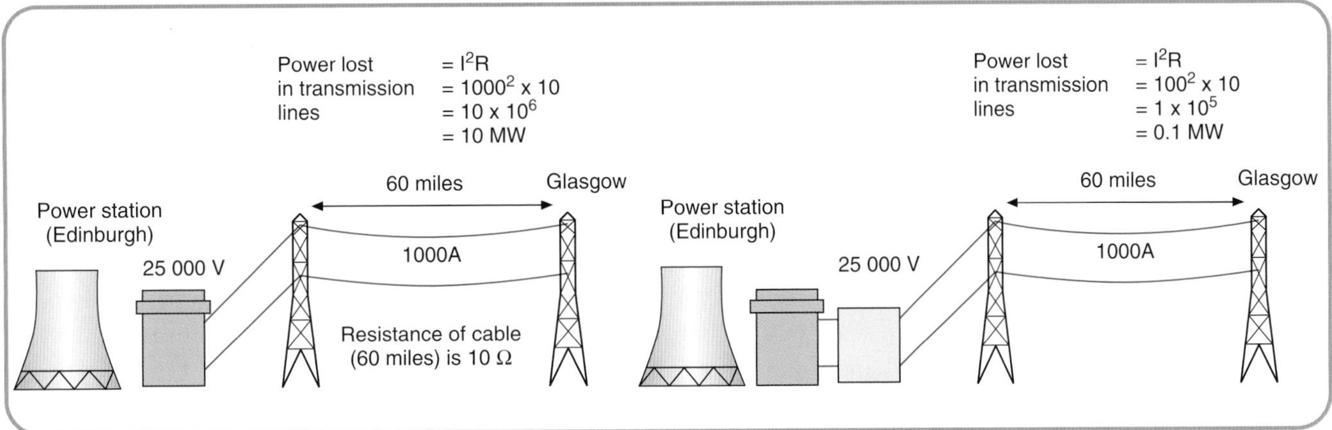

Figure 3.9 *Power loss in transmission lines can be greatly reduced by the use of transformers*

The transformer

A transformer consists of two separate coils of wire wound on the same iron core. The circuit symbol for a transformer is shown in Figure 3.10. The straight line between the coils represents the iron core.

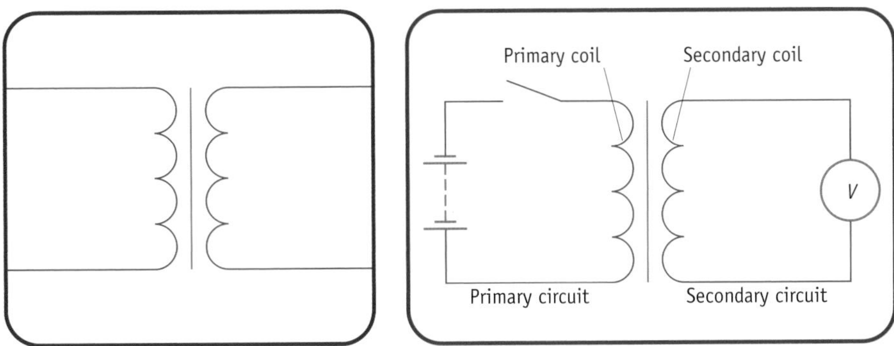

Figure 3.10 *Circuit symbol for a transformer*

Figure 3.11 *A transformer circuit*

Figure 3.11 shows a transformer connected to a switch and a battery. When the switch is closed, there is a current in the primary coil. This results in a magnetic field in the primary coil which rapidly builds up through both sets of coils, since they are joined by the iron core. This gives a changing magnetic field in the secondary coil and so a voltage (p.d.) is produced.

When the current in the primary circuit is steady, there is no change in the magnetic field and no voltage (p.d.) is produced in the secondary circuit. When the switch is opened, the primary current is switched off and the magnetic field around both the primary and secondary coils rapidly collapses (disappears). This changing magnetic field through the secondary coil results in a voltage (p.d.) being produced but in the opposite direction.

Removing the iron core through the coils decreases the effect since the iron core concentrates the magnetic field through the coils.

With a d.c. supply connected to the primary coil, a changing magnetic field can only be obtained by repeatedly opening and closing the switch. A more practical way of obtaining a changing magnetic field is to connect the primary coil to an a.c. supply. Now the current in the primary coil is always changing in size and direction. This means that the magnetic field is continually changing and so a voltage is continually produced.

Transformers therefore only work on a.c.

A transformer is used to investigate the relationship between the alternating voltages at the primary and secondary coils and the number of turns on the primary and secondary coils.

Primary turns (N_p)	Secondary turns (N_s)	N_s/N_p	Primary voltage (V_p)	Secondary voltage (V_s)	V_s/V_p
125	500	4	2 V	8 V	4
500	125	0.25	2 V	0.5 V	0.25
125	625	5	2 V	10 V	5
500	500	1	2 V	2 V	1
625	125	0.2	2 V	0.4 V	0.2

From this table we can conclude that:

$$\frac{N_s}{N_p} = \frac{V_s}{V_p}$$

- In a step-up transformer: secondary turns are more than the primary turns, i.e. $N_s > N_p$, thus $V_s > V_p$.
- In a step-down transformer: secondary turns are less than the primary turns, i.e. $N_s < N_p$, thus $V_s < V_p$.

Since the energy losses in transformers are normally very small (about 5% to heat, sound and leakage of the magnetic field) it is convenient to consider the transformer as being 100% efficient, i.e. an ideal transformer.

For an **ideal** transformer all the power at the primary is transferred to the secondary. Therefore:

input power = output power

$I_p \times V_p = I_s \times V_s$

i.e., $\dfrac{V_s}{V_p} = \dfrac{I_p}{I_s}$

Hence:

$$\frac{V_s}{V_p} = \frac{N_s}{N_p} = \frac{I_p}{I_s}$$

Example

The primary coil of an ideal transformer is connected to a 230 V mains supply. The output voltage from the transformer is 5 V. There are 400 turns on the secondary coil of the transformer. Calculate the number of turns on the primary coil.

Solution

$$\frac{N_p}{N_s} = \frac{V_p}{V_s}$$

$$\frac{N_p}{400} = \frac{230}{5}$$

$N_p = 46 \times 400 = 18{,}400$ turns

Example

An ideal transformer has 4000 turns on the primary coil and 100 turns on the secondary coil. The current in the primary coil is 30 mA. Calculate the current in the secondary coil.

Solution

$$\frac{I_s}{I_p} = \frac{N_p}{N_s}$$

$$\frac{I_s}{30 \times 10^{-3}} = \frac{4000}{100}$$

$I_s = 30 \times 10^{-3} \times 40$

$I_s = 1.2$ A

Physics facts and key equations for electromagnetism

- A magnetic field surrounds a current-carrying wire and is controlled by the current in the wire.
- A voltage can be produced in a coil when the magnetic field near the coil changes.
- The size of the voltage produced can be increased by increasing the strength of the magnetic field, increasing the number of turns on the coil and increasing the speed of movement of the coils through the magnetic field.
- A transformer consists of two separate coils of wire wound on an iron core.
- Transformers are used to change the size of an a.c. voltage.
- For a transformer, $\frac{V_p}{V_s} = \frac{N_p}{N_s}$.
- For a 100% efficient transformer,

$$\frac{V_p}{V_s} = \frac{N_p}{N_s} = \frac{I_s}{I_p}.$$

- Electrical power is distributed by the national grid system – the output voltage of 25,000 V from the power station is stepped-up by a transformer to 400,000 V for efficient transmission along the transmission lines. At the other end of the transmission lines a step-down transformer reduces the voltage to 230 V for use in our homes.

Questions

1 The diagram shown in Figure 3.12 shows some coils of wire connected to a voltmeter. When the coils are moved in the direction indicated a reading is obtained on the voltmeter. What change(s) could be made to the apparatus to increase the reading on the voltmeter?

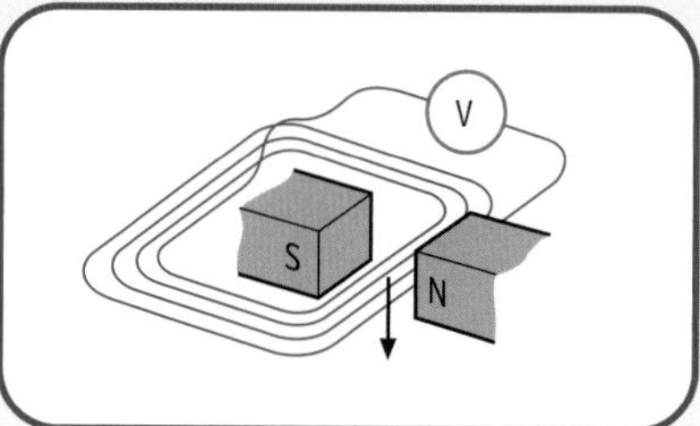

Figure 3.12

2 A 10 V a.c. supply is connected to the 40-turn primary coil of an ideal transformer. The voltage obtained from the secondary coil is 120 V. Calculate the number of turns on the secondary coil.

3 A 120 V a.c. supply is connected to the 8000-turn primary coil of an ideal transformer. A lamp is connected to the 400-turn secondary coil of the transformer. Calculate the potential difference across the lamp.

4 The output voltage from the 750-turn secondary coil of an ideal transformer is 5 V. The primary coil of the transformer has 1500 turns. Calculate the input voltage to the transformer.

5 An ideal transformer has a 250-turn primary coil and a 750-turn secondary coil. The current in the secondary coil is 1 A. Calculate the current in the primary coil.

6 A low-voltage heater is designed to work at 10 V and 2 A. An ideal transformer has 8280 turns on the primary coil. The primary coil is connected to a 230 V supply.
   a) Is the supply a.c. or d.c.?
   b) Calculate the current in the primary coil.
   c) Calculate the number of turns on the secondary coil.

4 Electronic components

At the end of this chapter you should be able to...

1. Give examples of output devices and the energy transformations involved.
2. Draw and identify the symbol for an LED.
3. State that an LED lights only when connected the correct way round.
4. Describe by means of a diagram a circuit which will allow an LED to light.
5. Calculate the value of the series resistor for an LED and explain the need for this resistor.
6. Give examples of input devices.
7. Describe the energy transformations involved in the following devices: microphone, thermocouple, solar cell.
8. State that, for most common thermistors, the resistance of the thermistor decreases as temperature increases.
9. State that the resistance of an LDR decreases with increasing light intensity.
10. Carry out calculations involving potential differenece, current and resistance for the thermistor and LDR.
11. Draw and identify the circuit symbol for an n-channel enhancement MOSFET.
12. Draw and identify the circuit symbol for an NPN transistor.
13. State that a transistor can be used as a switch.
14. Explain the operation of a simple transistor switching circuit.
15. Identify, from a list, devices in which amplifiers play an important part.
16. State that the output signal of an audio amplifier has the same frequency as, but a larger amplitude than, the input signal.
17. Carry out calculations involving input voltage, output voltage and voltage gain of an amplifier.

Output devices

Loudspeaker

A loudspeaker is connected to a signal generator. As the amplitude (energy of the electrical signal) from the signal generator is increased the sound from the loudspeaker gets louder. A loudspeaker changes electrical energy into sound.

Electric motor

An electric motor is connected to a variable d.c. supply voltage. The speed of the motor increases as the voltage (p.d.) is increased. An electric motor changes electrical energy into kinetic energy. Reversing the connections to the supply reverses the direction of rotation of the motor.

Relay

A relay is a switch operated by an electromagnet. When switch S is closed the current in the relay produces a magnetic field which causes the relay switch to close. This completes the lower electrical circuit and the lamp lights. When S is opened there is no current in the relay. The relay switch opens and the lamp goes out. The relay changes electrical energy into kinetic energy i.e. the opening or closing of a switch.

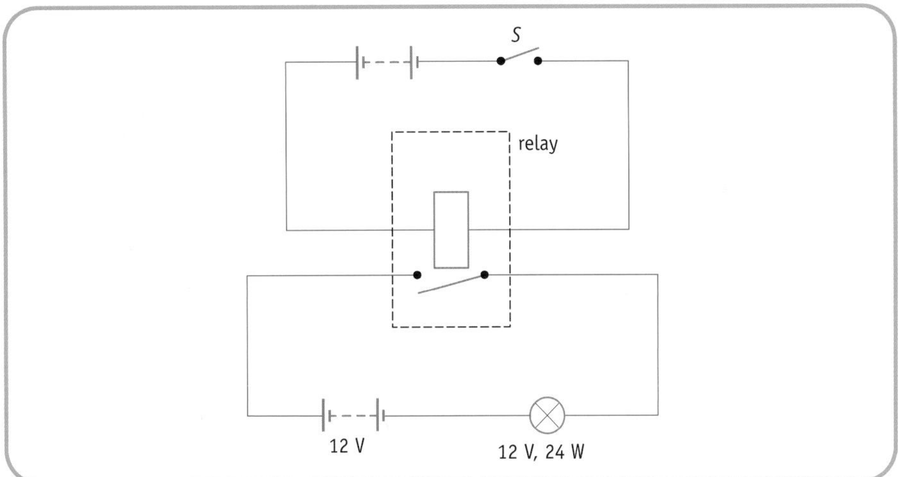

Figure 4.1 *A relay circuit – closing switch S allows the lamp to light*

Filament lamp

A filament lamp consists of a thin tungsten wire (filament) in a glass container. When there is an electric current in the wire, electrical energy is changed into heat and light in the filament. A lamp is connected to a variable d.c. voltage supply. Increasing the voltage (p.d.) across the lamp increases the current in it and so the lamp gets brighter. No difference is observed when the connections from the supply to the lamp are reversed. The filament in the lamp requires a relatively large current to light properly and gets very hot in operation.

Light-emitting diode (LED)

Light-emitting diodes are made by joining two special materials together to produce a junction. When there is an electric current in the junction, electrical energy is changed into light. However, too large a current – or indeed too high a p.d.– will destroy the junction. To prevent this a resistor must be connected in series with the LED. Figure 4.2 shows an LED and a resistor connected to a variable d.c. voltage supply. Increasing the voltage (p.d.) across the LED increases its brightness. The LED does not light when the connections from the supply are reversed. The LED only requires a small current to light and does not get hot in operation. LEDs are available in red, orange, yellow, green, blue and white colours.

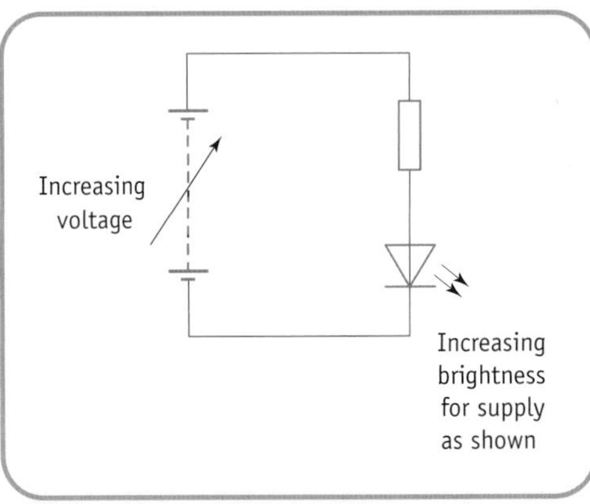

Figure 4.2 *A variable voltage supply connected to a resistor and an LED*

Example

The maximum voltage allowed across an LED is 1.8 V and the current in it must not exceed 10 mA. The LED is connected to a 6.0 V d.c. supply. Calculate the value of the resistor, *R*, connected, in series, with the LED.

Solution

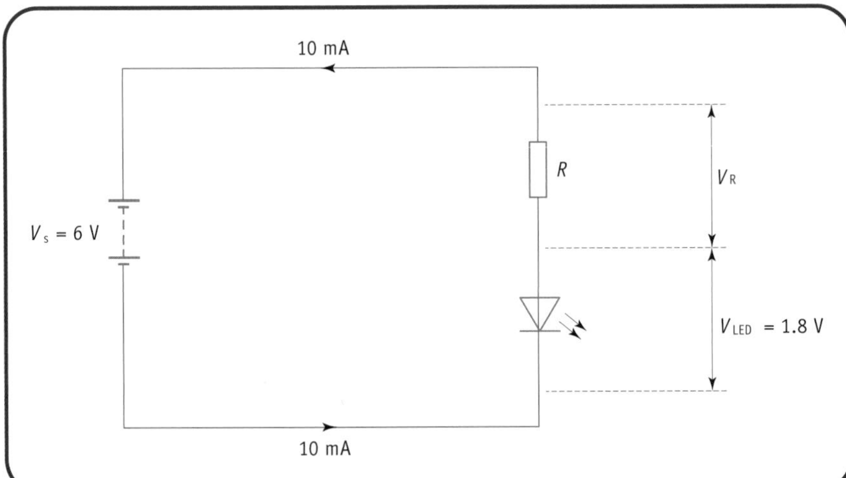

Figure 4.3 *Circuit diagram*

Since the LED and resistor are connected in series then $V_s = V_{LED} + V_R$ and the current in both components is the same.

Therefore $V_R = V_s - V_{LED} = 6.0 - 1.8 = 4.2$ V

$$V_R = IR$$

$$4.2 = 10 \times 10^{-3} \times R$$

$$R = \frac{4.2}{10 \times 10^{-3}} = 420\ \Omega$$

Did you know?

Light emitting diode (LED) lamps – a large number of LEDs in the form of a lamp – are replacing filament lamps in traffic lights and side and brake lamps for some buses and cars. LED lamps consist of a cluster of ultra-bright LEDs. LED lamps have a much lower power consumption, longer life, and are more reliable and brighter than filament lamps.

Figure 4.4 *LED lamps are used to light this set of traffic lights*

Input devices

Microphone

A microphone is connected to an oscilloscope. As louder notes are played into the microphone, the amplitude of the trace displayed on the oscilloscope increases. A microphone changes sound into electrical energy. The louder the sound, the greater the electrical energy produced.

Thermocouple

A thermocouple is composed of two different types of wire joined together. A thermocouple is connected to a voltmeter as shown in Figure 4.5. When the junction of the thermocouple is placed in a Bunsen flame, the voltmeter reading increases. A thermocouple changes heat into electrical energy. The higher the temperature of the junction, the greater the electrical energy produced.

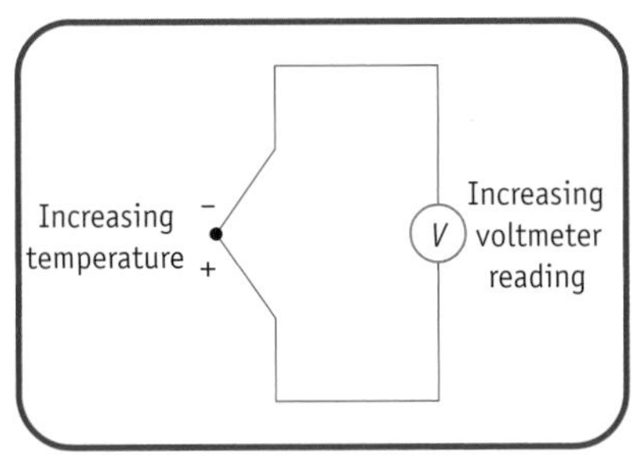

Figure 4.5 *A thermocouple connected to a voltmeter*

Solar cell

A solar cell is connected to a voltmeter as shown in Figure 4.6. When the solar cell is exposed to more light, the voltmeter reading increases. A solar cell

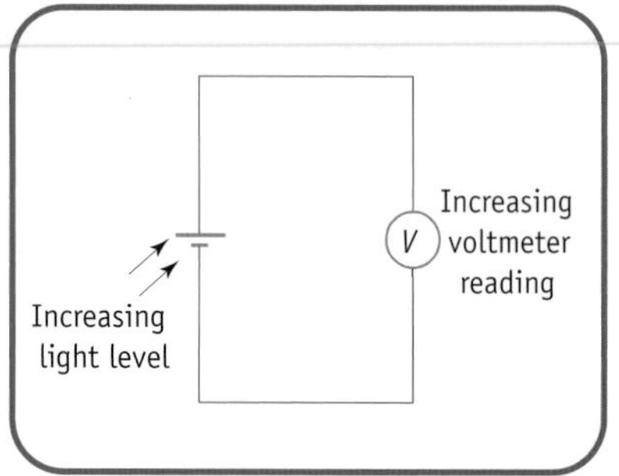

Figure 4.6 *A solar cell connected to a voltmeter*

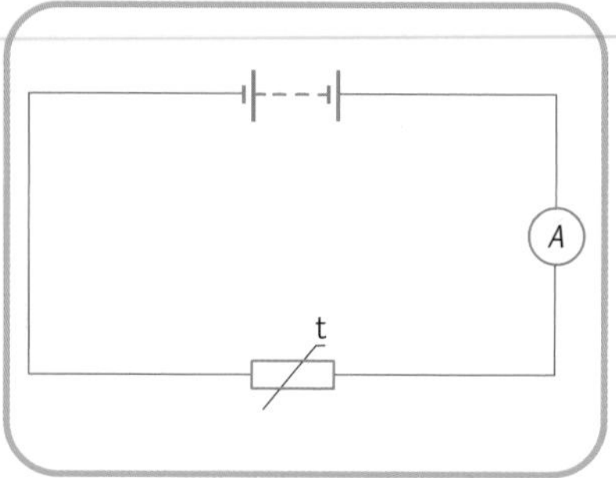

Figure 4.7 *A thermistor circuit*

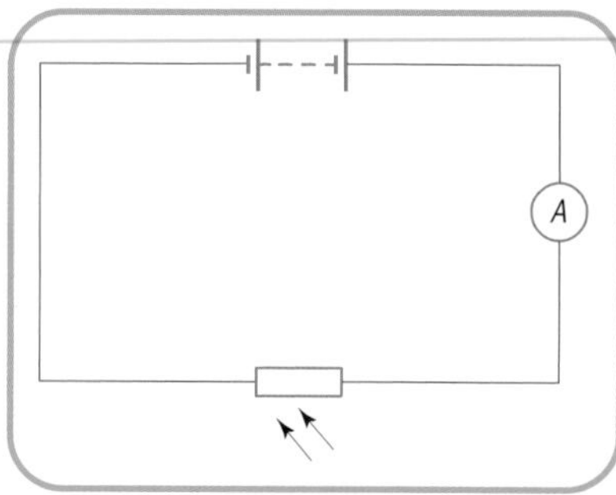

Figure 4.8 *An LDR circuit*

changes light into electrical energy. The brighter the light shining on the solar cell, the greater the electrical energy produced.

Thermistor

When the thermistor shown in Figure 4.7 is heated, the ammeter reading increases – therefore the resistance of the thermistor must be decreasing.

The resistance of most thermistors usually decreases with increasing temperature.

temperature ↑ – resistance of thermistor ↓

Light-dependent resistor (LDR)

When the light-dependent resistor, shown in Figure 4.8, is exposed to brighter light, the ammeter reading increases – therefore the resistance of the LDR must be decreasing.

As the light gets brighter (light intensity increases) the resistance of the LDR decreases.

light intensity ↑ – resistance of LDR ↓

Example

A thermistor, 1 kΩ resistor and an ammeter are connected in series to a 6 V supply as shown in Figure 4.9. When the thermistor is at a temperature of 20°C the reading on the ammeter is 2 mA.

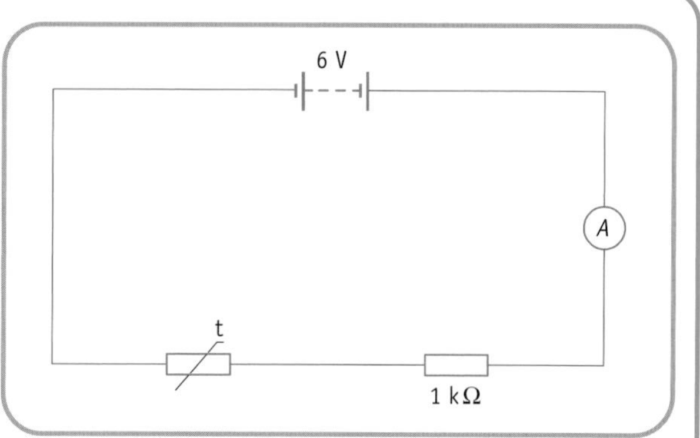

Figure 4.9 *Example of a thermistor*

a) Calculate the voltage across the 1 kΩ resistor at 20°C.
b) Find the resistance of the thermistor at 20°C.

Solution

a) Voltage across resistor, $V_R = IR_R = 2 \times 10^{-3} \times 1 \times 10^3 = 2$ V

b) $V_S = V_R + V_{thermistor}$

$6 = 2 + V_{thermistor}$

$V_{thermistor} = 4$ V

But $V_{thermistor} = IR_{thermistor}$

$4 = 2 \times 10^{-3} \times R_{thermistor}$

$$R_{thermistor} = \frac{4}{2 \times 10^{-3}} = 2000\ \Omega$$

Example

A student uses the apparatus shown in Figure 4.10 to measure the resistance of an LDR. When light is shone on the LDR the reading on the voltmeter is 2 V and the reading on the ammeter is 8 mA. Calculate the resistance of the LDR at this light level.

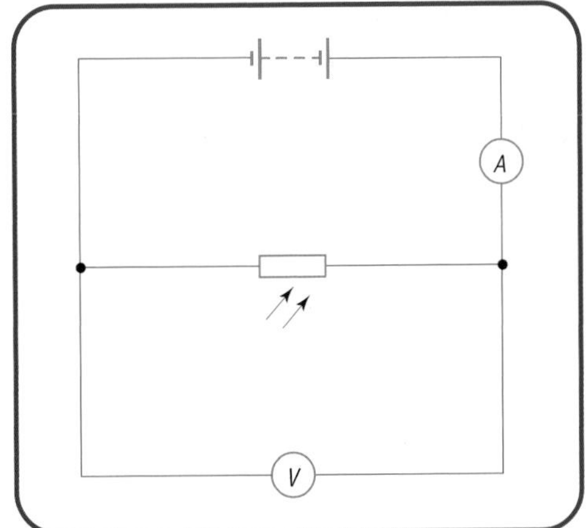

Figure 4.10 *Example of an LDR*

Solution

$V_{LDR} = IR_{LDR}$

$2 = 8 \times 10^{-3} \times R_{LDR}$

$$R_{LDR} = \frac{2}{8 \times 10^{-3}} = 250\ \Omega$$

Transistors

- A MOSFET (metal oxide semiconductor field effect transistor) – this has three terminals, called the gate, the source and the drain. The symbol for an n-channel MOSFET is shown in Figure 4.11.
- NPN transistor – this has three terminals, called the base, the emitter and the collector. The symbol for a NPN transistor is shown in Figure 4.12.

The MOSFET and the NPN transistor are both types of electronic switch with no moving parts.

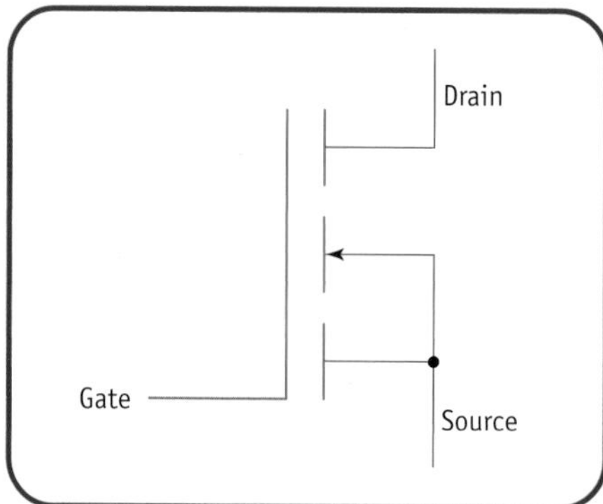

Figure 4.11 *MOSFET symbol*

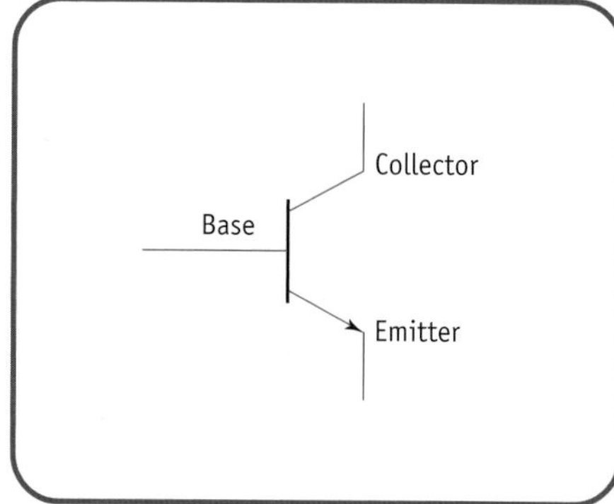

Figure 4.12 *NPN transistor symbol*

The switching of an NPN transistor (MOSFETs operate in a similar way) is controlled by the voltage (p.d.) applied to the emitter-base. The transistor is off (non-conducting) when the emitter-base voltage (p.d.) is below a certain value – i.e. the electronic switch is open. However, the transistor is on (conducting) when the emitter-base voltage (p.d.) is equal to or above this certain value, i.e. the electronic switch is closed.

A temperature-controlled circuit

Figures 4.13 and 4.14 show two temperature controlled circuits. The variable resistor, in each circuit, is adjusted until at room temperature the LED is just off.

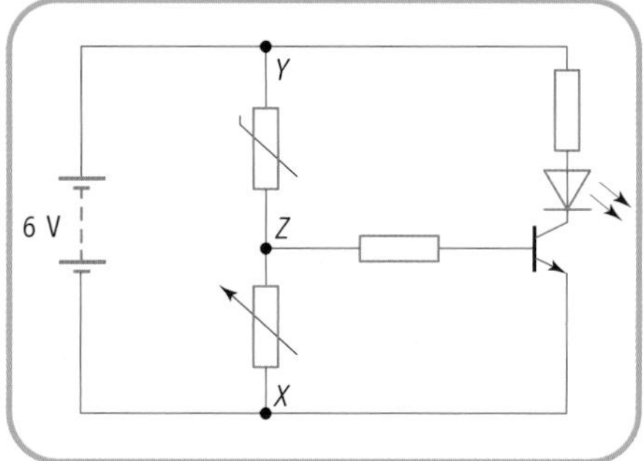

Figure 4.13 *A temperature-controlled circuit*

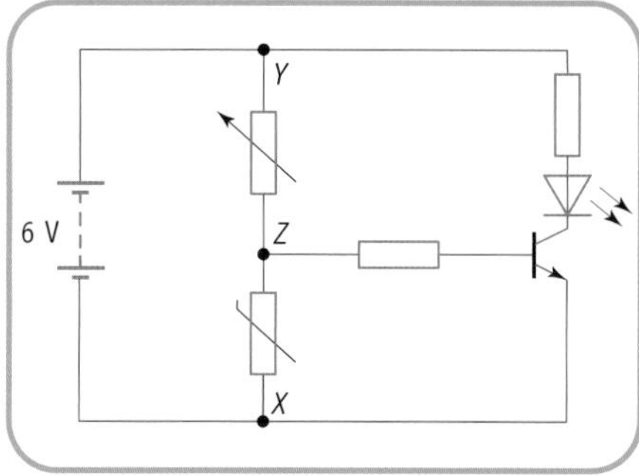

Figure 4.14 *Alternative temperature-controlled circuit*

LED is off
Heat thermistor
Temperature of thermistor ↑
$R_{thermistor}$ ↓
$V_{thermistor}$ ↓ so V_{zy} ↓
V_{xz} ↑
Transistor switches on
LED lights

LED is off
Cool thermistor
Temperature of thermistor ↓
$R_{thermistor}$ ↑
V_{xz} ↑
Transistor switches on
LED lights

Note. A variable resistor is used in this type of circuit instead of a fixed resistor. The variable resistor allows the circuit to be adjusted to different conditions (temperature in this case) before the output device comes on (or goes off).

A light-controlled circuit

Figures 4.15 and 4.16 show two light-controlled circuits. The variable resistor is adjusted, in each circuit, until at normal light level the LED is just off.

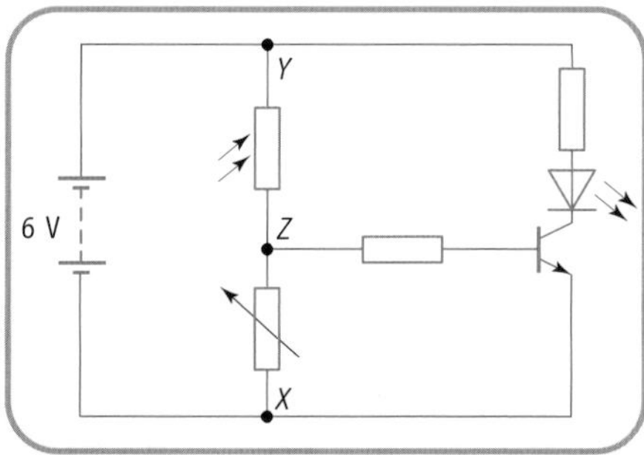

Figure 4.15 *A light-controlled circuit*

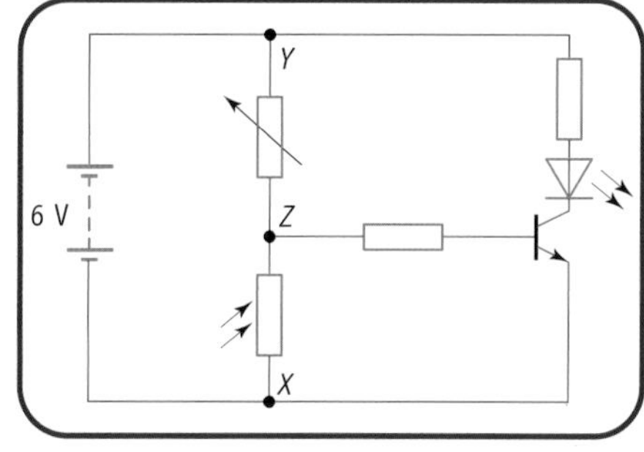

Figure 4.16 *Alternative light-controlled circuit*

LED is off
Shine more light on LDR
R_{LDR} ↓
V_{LDR} ↓ so V_{ZY} ↓
V_{xz} ↑
Transistor switches on
LED lights

LED is off
Cover LDR
R_{LDR} ↑
V_{XZ} ↑
Transistor switches on
LED lights

Amplifiers

An amplifier is a device which is generally used to make electrical signals larger. Figure 4.17 shows an amplifier being used to make the electrical signal from a signal generator larger. The traces produced by both input and output signals from the amplifier are displayed on the screens of identical oscilloscopes as shown in Figure 4.18. The amplitude of the trace of the output signal from the amplifier is larger than the amplitude of the trace of the input signal – the extra energy comes from the electrical supply to the amplifier. The frequency of the output from the amplifier is the same as the frequency of the input signal. In most audio devices such as radios, televisions and hi-fis the amplifier is the volume control.

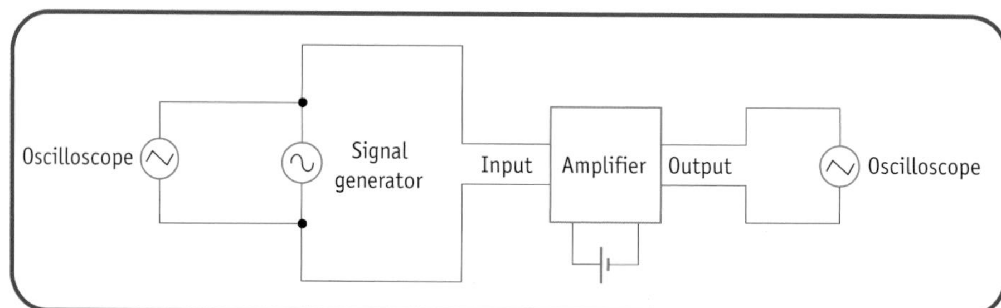

Figure 4.17 *An amplifier circuit*

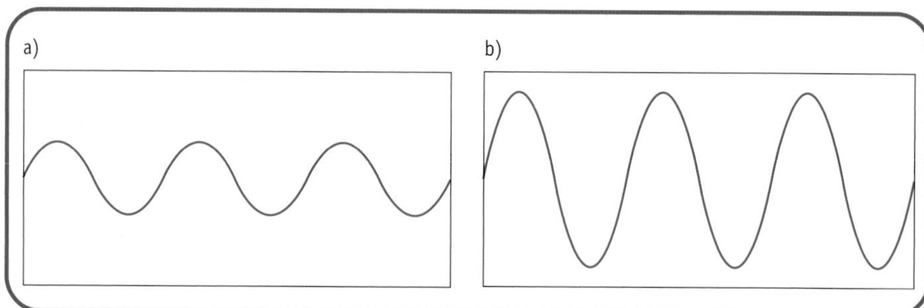

Figure 4.18 *a) Input voltage to amplifier b) Output voltage from amplifier*

Figure 4.19 *A hi-fi contains an amplifier*

Voltage gain

An amplifier has an input voltage of 0.5 V and an output voltage of 5.0 V – the output voltage is 10 times larger than the input voltage, i.e. the voltage gain is 10:

$$\text{Voltage gain} = \frac{\text{output voltage}}{\text{input voltage}}$$

Note that voltage gain does not have a unit.

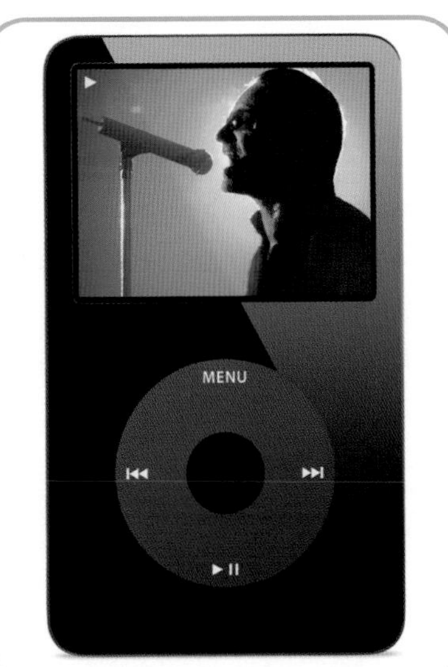

Figure 4.20 *Amplifiers are to be found in most modern electronic devices such as this iPod.*

Example

The voltage gain of an amplifier is 200. The voltage at the input of the amplifier is 20 mV. Calculate the voltage at the output of the amplifier.

Solution

$$\text{Voltage gain} = \frac{\text{output voltage}}{\text{input voltage}}$$

$$200 = \frac{\text{output voltage}}{20 \times 10^{-3}}$$

$$\text{output voltage} = 200 \times 20 \times 10^{-3} = 4\ \text{V}$$

Physics facts and key equations for electrical components

- An output device changes electrical energy into some other form of energy.
- There are a number of output devices: examples are loudspeaker, relay, filament lamp and light-emitting diode (LED).
- LEDs will only light up when connected to the power supply the correct way round.
- A resistor should always be connected in series with an LED – this protects the LED from damage from too high a current in it (or too high a voltage across it).
- Most input devices change some form of energy into electrical energy.
- There are a number of input devices: examples are microphone, thermocouple, solar cell, thermistor and light-dependent resistor (LDR).
- The resistance of a thermistor changes with temperature – the resistance of most thermistors decreases as the temperature increases.
- The resistance of an LDR decreases with increasing light level.
- An NPN transistor and a MOSFET are electrically operated switches.
- A transistor is non-conducting (OFF) for voltages below a certain value but conducting (ON) at voltages at or above this certain value.
- An amplifier is a device which makes electrical signals larger.
- Audio amplifiers are found in devices such as radios, televisions, hi-fis, intercoms and loudhailers.
- The output signal from an audio amplifier has the same frequency as, but a larger amplitude than, the input signal.
- For an amplifier:

$$\text{voltage gain} = \frac{\text{output voltage}}{\text{input voltage}}$$

Questions

1 Name two output devices and state the energy conversion for each one.

2 A student designs a suitable circuit to light an LED. The student uses the following components: a 6 V battery, a switch, an LED and a resistor.

a) Draw a suitable diagram, which will allow the LED to light when the switch is closed.

b) The maximum voltage across the LED must not exceed 1.75 V. The maximum current in the LED must not exceed 11 mA. Calculate the value of the resistor required for the circuit.

3 State the energy conversion for: (a) a microphone, (b) a thermocouple, (c) a solar cell.

4 A thermistor is connected to a power supply. A student connects an ammeter and a voltmeter in the circuit so that she can find the resistance of the thermistor. When the temperature of the thermistor is 18°C the reading on the voltmeter is 6 V and the reading on the ammeter is 12 mA.

a) Calculate the resistance of the thermistor at this temperature.

b) The temperature of the thermistor rises to 20°C. The reading on the voltmeter remains at 6 V. Suggest a value for the reading on the ammeter.

5 Figure 4.21 shows a circuit built by a student to detect when the element of an electric cooker is hot. The thermistor is placed near to the element. The resistance of the thermistor decreases as its temperature increases. Explain what will happen as the temperature of the cooker element increases.

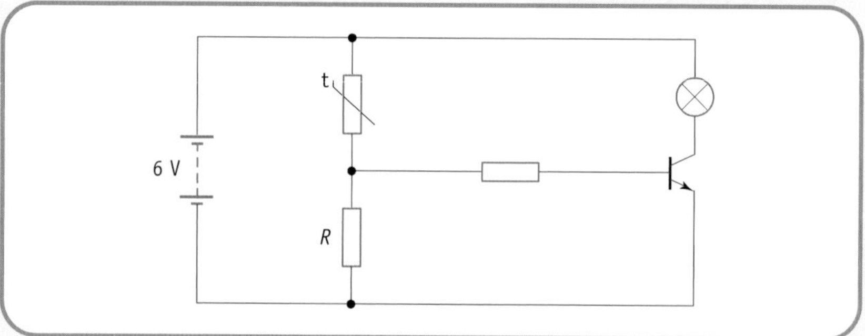

Figure 4.21

6 Draw the circuit symbol for: (a) an NPN transistor, (b) a MOSFET transistor.

7 The input voltage to an audio amplifier is 10 mV and the output voltage from it is 3.0 V.

a) Calculate the voltage gain of this amplifier.

b) How do the frequencies of the input and output signals of an amplifier compare?

8 The voltage gain of an amplifier is 200. Calculate the voltage at the output of the amplifier when the voltage at its input is 3 mV.

Exam Questions

1 a) Two resistors are connected in series to a 9 V supply as shown in Figure E.1.

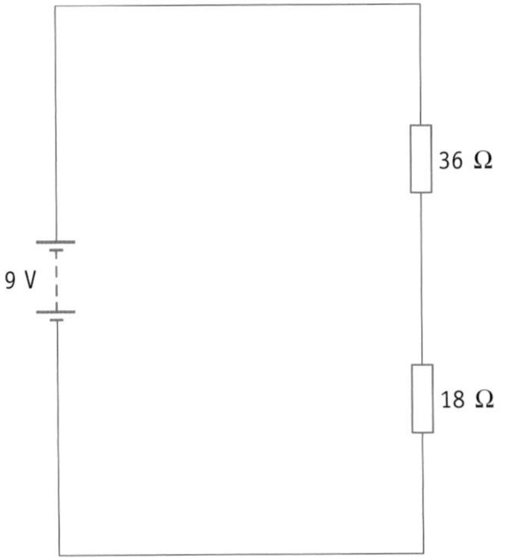

Figure E.1

(i) Calculate the current drawn from the supply.
(ii) Find the p.d. across the 18 Ω resistor.

b) An electrical appliance has three resistors connected as shown in Figure E.2.
(i) Calculate the resistance between points *A* and *B*.
(ii) Calculate the power rating of the appliance when points *A* and *B* are connected to a 230 V mains supply.

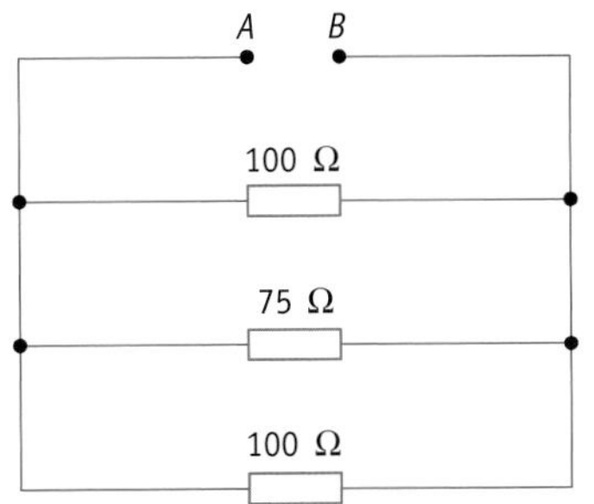

Figure E.2

2 A set of Christmas tree lights consists of 19 lamps and a resistor, all connected in series. The lamps are labelled 12 V, 3 W. The set is connected to the 230 V mains supply and switched on. The lamps are working at their rated voltage.

a) Why must the 19 lamps and resistor be connected in series?
b) Calculate the current in each of the lamps.
c) Show that the resistance of the resistor is 8 Ω.
d) How much charge flows through each lamp in 1 minute?
e) Calculate the resistance of each lamp.

3 The element of an electric kettle has a power rating of 2116 W. The element is connected to a 230 V mains supply and switched on for 180 s.

a) (i) How much electrical energy is transferred by the element into heat in 180 s?
(ii) Where does this energy change occur?
b) Calculate the resistance of the element when it is switched on.

4 An electrical component and a resistor are connected in series with a 10 V supply as shown in Figure E.3.

a) Name component *X*.
b) Explain why resistor *R* is necessary in the circuit.
c) The voltage across component *X* is 1.8 V and the current in the circuit 11 mA. Calculate the value of resistor *R*.

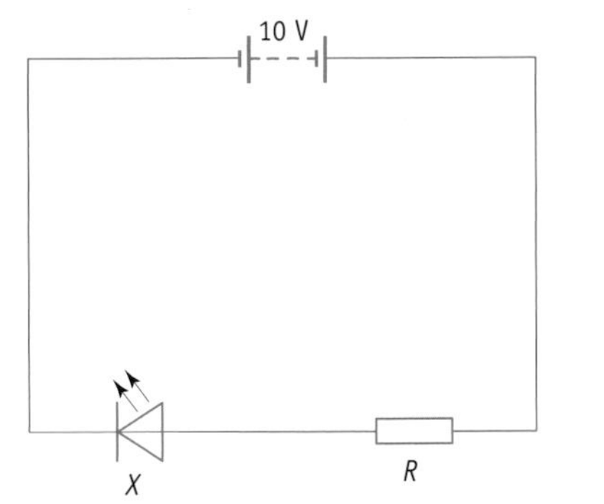

Figure E.3

5 An ideal transformer has a 2000-turn primary coil and a 200-turn secondary coil. A 12 V, 24 W lamp is connected to the secondary coil of the transformer. The lamp is working at its rated voltage.

a) Calculate the input voltage to the primary coil of the transformer.
b) Find the current in the lamp.
c) Find the current in the primary coil of the transformer.

6 An electric cooker has two heating plates. Each heating plate is made up of two heating elements, each of resistance 100 Ω. The heating elements are connected in the circuits as shown below.

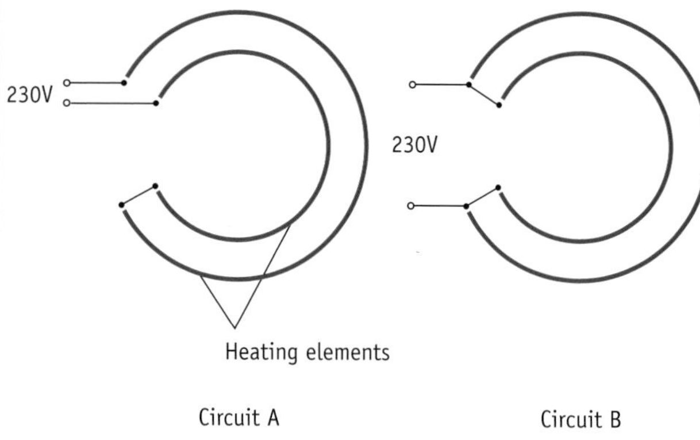

Figure E.4

a) Which circuit A or B, shows the heating elements connected in series?
b) Calculate the resistance of (i) circuit A, (ii) circuit B.
c) Both circuits are connected to a 230 V mains supply and switched on. Calculate the current in (i) circuit A, (ii) circuit B.
d) Which circuit A or B, would heat a pot of soup in the shortest time? You must show the calculations you use to arrive at your answer.

7 In the circuit shown below, the variable resistor is adjusted so that the LED is just off when the temperature of the thermistor is 20°C.

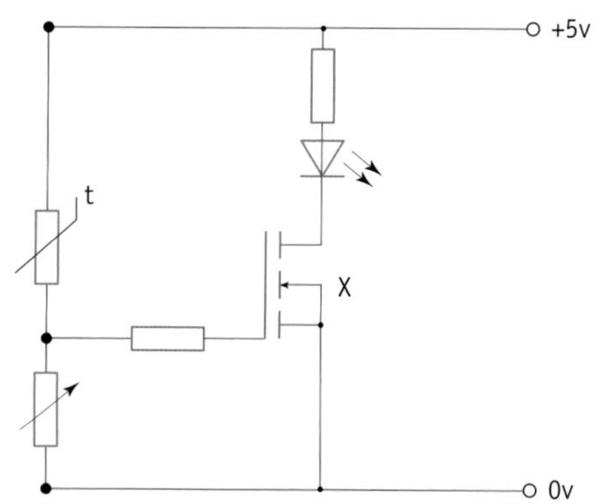

Figure E.5

a) Name component X.
b) The resistance of the thermistor decreases as its temperature increases. Explain why the LED lights when the temperature of the thermistor rises to 22°C.
c) The circuit could be altered to warn a gardener of low temperture conditions. What alteration should be made to the circuit diagram that would allow the LED to light during low temperatures?

Unit 2

Mechanics and Heat

Understanding the mechanics of objects such as cars and aeroplanes can help us to travel safely, conserve energy and understand cause and effect. This unit examines calculating and measuring forces and motions, including speed, friction, collision and changes in temperature.

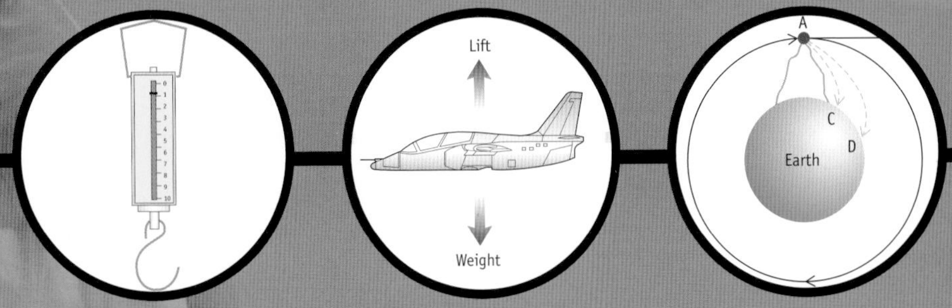

5 Kinematics

At the end of this chapter you should be able to...

1 Describe how to measure an average speed.
2 Carry out calculations involving the relationship between distance, time and average speed.
3 Describe how to measure instantaneous speeds.
4 Identify situations where average and instantaneous speed are different.
5 Describe what is meant by vector and scalar quantities.
6 State the difference between distance and displacement.
7 State the difference between speed and velocity.
8 Explain the terms speed, velocity and acceleration.
9 State that acceleration is change in velocity per unit time.
10 Draw velocity–time graphs involving more than one constant acceleration.
11 Describe the motions represented by a velocity–time graph.
12 Calculate displacement and acceleration, from velocity–time graphs, for more than one constant acceleration.
13 Carry out calculations involving the relationships between initial velocity, final velocity, time and uniform acceleration.

Speed

Speed is the distance travelled by a vehicle or a person in one second. Your speed will vary throughout a journey and we will talk about average speed:

$$\text{average speed} = \frac{\text{distance travelled}}{\text{time taken}}$$

In symbols, $v = \dfrac{d}{t}$

Average speed is the constant speed needed to cover the distance travelled in the time allowed. Speed and average speed have the same unit of m/s.

Measuring average speed

Using a stopwatch, trolley and measuring tape

The distance is measured between two marked points a few metres apart on the ground, using a measuring tape. The time for the journey is measured by starting a stopwatch when the trolley reaches the first point (x) and stopping the watch when it reaches the second point (y) (Figure 5.1). The speed can change several times during the journey and this is why it is an average speed.

$$\text{average speed} = \frac{\text{distance}}{\text{time}} = \frac{\text{d}}{\text{t}}$$

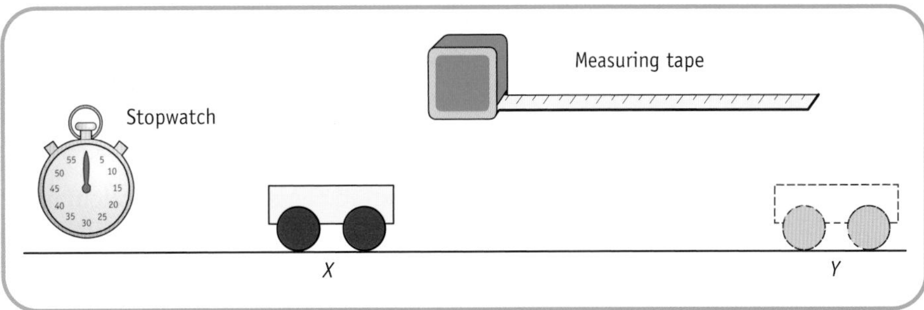

Figure 5.1 *Measuring average speed*

Instantaneous speed

This is the speed at one instant of time – for example, just as a runner crosses a finishing line or a car passes a certain point on a racing track. If we could measure very small time intervals such as thousandths of a second then we could measure the speed of any object just at that moment in time.

Instantaneous speed is the speed of an object at a particular time (instant).

$$\text{average speed} = \frac{\text{total distance travelled}}{\text{total time taken}} = \frac{d}{t}$$

When the time taken (t) is very small, the closer the average speed is to the instantaneous (actual) speed.

Example

A car moves a distance of 150 m in a time of 7.5 s. Calculate the car's average speed.

Solution

$$\text{average speed} = \frac{150}{7.5} = 20 \text{ m/s}$$

This time interval is too large to give a reasonable estimate of the instantaneous speed. If we use a much smaller distance and use a much smaller time interval we can obtain a measure of the instantaneous speed.

distance = 1.0 m

time taken = 0.04 s

$$\text{average speed} = \frac{1.0}{0.04} = 25 \text{ m/s}$$

This is closer to the instantaneous speed as the time interval involved is very small.

This means that instantaneous speed = average speed of an object, **provided the time used is very small**. However, this small time interval is very difficult to measure since human reaction time is involved. This is the time to operate a stopwatch. To avoid this problem an electronic system of time measurement is required which can operate without reaction time.

Measuring instantaneous speed using a computer

The computer uses an internal clock, which allows very small time intervals to be measured. This allows the calculation of instantaneous speed if a distance is measured. The computer needs to be connected to a piece of equipment called a light gate.

Figure 5.2 *Measuring instantaneous speed*

The computer starts timing when a light beam in the light gate is cut by a card, and stops when the light beam is restored. The time taken for the card to pass through the beam is recorded in the computer (Figure 5.2).

The computer has the information about the length of the card and calculates the speed of the vehicle using:

$$\text{speed} = \frac{\text{length of card}}{\text{time on computer}}$$

Typical results might be:

Length of card = 5 cm = 0.05 m

Time measured by computer= 0.02s

$$\text{Speed} = \frac{0.05}{0.02}$$

$$= 2.5 \text{ m/s}$$

The police measure both average speed and instantaneous speed.

Figure 5.3 *Speed camera and road markings*

Average speed can be found from the speed cameras, which time between two marks on the ground. Since the distance between the marks is known, the average speed can be calculated (Figure 5.3).

The distance between the marks is 2 m. The camera takes pictures 0.5 s apart. The number of marks between each picture are noted and the speed calculated.

In a similar way the single speed camera can be used. This uses a sensor to detect when a car is possibly speeding. The camera takes two pictures which are 0.5 seconds apart. The road surface has markings which are 2 m apart, allowing the speed to be found (Figure 5.4).

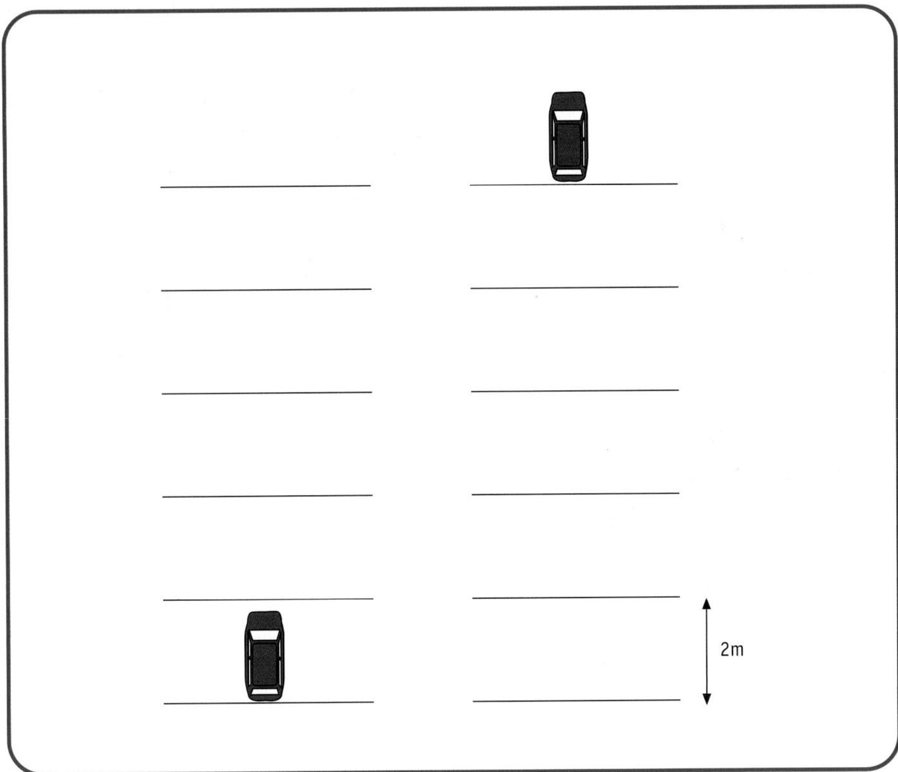

Figure 5.4

Example

A speed camera takes two photographs (represented in Figure 5.4), 0.5 s apart. The distance between each mark is 2 metres. The speed limit is 18 m/s. Show by calculation if the driver was speeding (Figure 5.5).

Solution

The car has travelled 5 complete spaces.

This means the car has travelled 10 m in a time of 0.5 s.

$$\text{Average speed} = \frac{10}{0.5} = 20 \text{ m/s}$$

The driver of the car was breaking the speed limit.

Speed Guns

The speed gun which is used to check speeds measures an instantaneous speed. High-frequency waves are sent out from the gun and return from a moving object at a different frequency. If an emergency vehicle passes at high speed with its siren sounding then the pitch or frequency of the note will change. The difference in frequency can be used to calculate the speed (Figure 5.5).

Figure 5.5 *A speed gun*

Vectors and scalars

When you travel in a car or an aircraft, the driver or pilot can tell you the speed at which the vehicle is travelling. This tells you nothing about the direction of travel. If you specify the direction then you have a vector.

A scalar has only magnitude or size. Examples are mass, distance, speed, temperature, energy and volume. These quantities can be added in the normal way using arithmetic: 5 kg of potatoes and 3 kg of potatoes give us 8 kg of potatoes!

Vectors have magnitude and direction. Vectors must be added using the ideas of a scale diagram since vectors are represented by straight lines drawn to scale. Examples of vectors are displacement and velocity.

Distance is how far you travel. Displacement is the shortest distance from start to finish. It needs a magnitude and a direction to describe it fully. Examples are 2 km due west; 500 m at a bearing of $(023)^\circ$. The direction can be described as a three figure bearing measured clockwise from north or as a two figure bearing from the points of the compass, for example 23° east of north. It is the vector showing the distance and direction from your starting point to your finishing point in a straight line.

Example

Lesley walks 7 km south and then 3 km north. What is the displacement from the starting point?

Solution

The two lines are shown in Figure 5.6.

- The lines are shown with an arrow to indicate direction and the tail of one vector should join at the head of the other.
- The vectors should be drawn to scale and the scale stated.
- A line from the tail of the first vector to the head of the second vector represents the resultant vector.
- The final or resultant displacement is 4 km south.

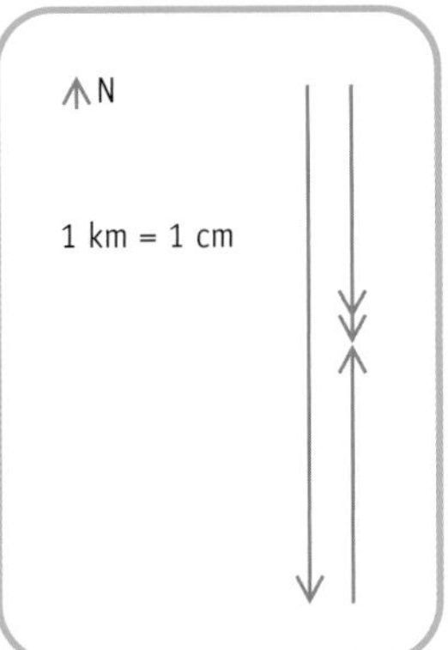

Figure 5.6

Example

If Gordon walks 20 m north and then 30 m east then he has walked 50 m (Figure 5.7). With a vector diagram drawn as shown his resultant displacement can be measured or calculated. His resultant displacement from his starting position is 36 m at (056)°

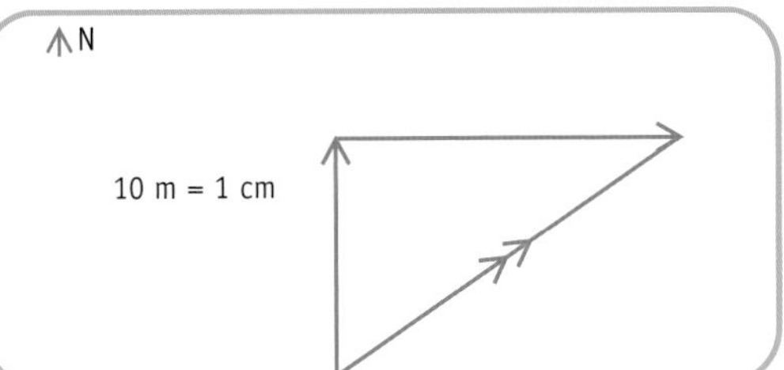

Figure 5.7

Similarly velocity is a vector since it is speed with a direction:

$$\text{average velocity} = \frac{\text{displacement}}{\text{time}}$$

Example

An aircraft is travelling east at 140 m/s . A wind is blowing north at 20 m/s. What is the velocity of the aircraft relative to the ground?

Solution

The two vectors for the aircraft and the wind velocity are shown in Figure 5.8. The resultant velocity can be found by a scale diagram or by pythagoras and is 141.4 m/s

The angle is 8.1° N of E.

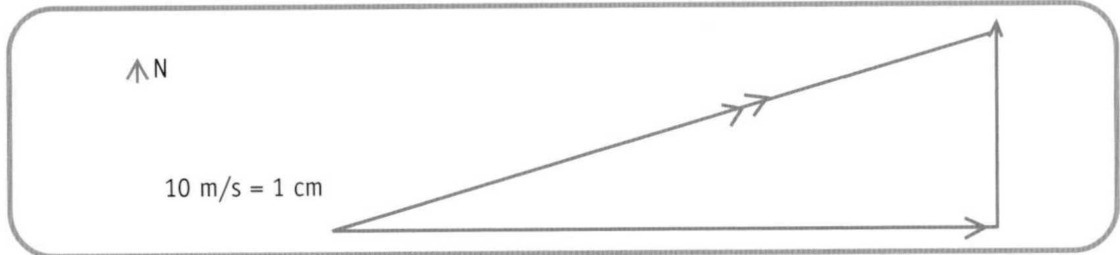

Figure 5.8

Acceleration

The table below gives data for a number of cars.

Car model	Engine size (litres)	Miles per gallon	Maximum Speed (mph)	Time from 0 to 60 mph (s)
Audi A3	1.6	29	116	11.9
BMW 1	1.16	27	124	10.8
Citroen C3	1.1	36	95	15.5
Fiat Punto	1.2	33	96	14.3
Ford Focus	1.4	32	102	14.1
Honda Civic	1.4	35	109	12.1
Jaguar X type	2.0	36	120	11.4
Mercedes A class	1.5	36	109	12.6
Nissan Almera	1.5	33	110	13.1
Peugeot 307	1.4	33	107	12.8
Renault Megane	1.4	31	114	12.5
Toyota Corolla	1.4	34	115	12.0
Vauxhall Astra	1.4	35	111	13.7
Volkswagon Golf	1.4	29	102	14.7

The engine size differs from one car to the next and the cost is not shown. The main part of the table that is relevant is the last column, which is the time taken to go from 0 to 60 mph. This allows us to compare different cars.

Acceleration (a) is the change in velocity of an object in one second:

For the Ford Focus the acceleration is $\frac{60 - 0}{14.1} = 4.26$ mph per second

This means that every second the Focus increases its speed by 4.26 mph if it is travelling in a straight line.

The table shows the speed of the Ford Focus at one second intervals.

Time (s)	Speed (mph)
0	0
1	4.26
2	8.52
3	12.78
4	17.04
5	21.3

Acceleration can be defined precisely as:

$$\text{acceleration} = \frac{\text{change in velocity}}{\text{time for change}} = \frac{\text{final velocity} - \text{initial velocity}}{\text{time for change}}$$

$$a = \frac{v - u}{t}$$

where v = final velocity, u = initial velocity and t = time for change in velocity to occur.

Acceleration is usually calculated in metres per second per second. We write this as m/s^2.

A car is travelling in a straight line with a constant acceleration of 5 m/s^2. This means that the velocity of the car increases by 5 m/s in every second. If the car starts from rest then:

after 1 s, velocity = 5 m/s

after 2 s, velocity = 10 m/s

after 3 s, velocity = 15 m/s

after 4 s, velocity = 20 m/s

Example
A cheetah is one of the fastest animals in the world. Starting from rest it accelerates uniformly and can reach a velocity of 24 m/s in 3 seconds. What is its acceleration?

Solution

$u = 0$; $v = 24$ m/s; $a = ?$; $t = 3$ s

$$\text{Acceleration} = \frac{v - u}{t} = \frac{24 - 0}{3} = \frac{24}{3} = 8 \text{ m/s}^2$$

When an object is slowing down it will have a negative value for its acceleration – this is called a deceleration.

Example
A student on a scooter is travelling on a straight road and slows down uniformly from 6 m/s to 2m/s in 4 seconds. Calculate the deceleration of the scooter.

Solution

$u = 6$ m/s; $v = 2$ m/s; $a = ?$; $t = 4$ s

$$\text{Acceleration} = \frac{v - u}{t} = \frac{2 - 6}{4} = \frac{-4}{4} = -1 \text{ m/s}^2$$

Deceleration = 1 m/s^2

Did you know?

Recently a newspaper decided to see which was the fastest car in the world in terms of acceleration. They tested three cars. The results for 0 to 60 mph are shown below.

Caterham 7	average time 3.28 s
Atom	average time 2.91 s
Noble	average time 3.65 s

It would appear that the Atom had won, but the tyres used were racing tyres not normal road tyres.

By comparison the McLaren F1 11 racing car takes 3.2 s to go from 0 to 60 mph.

Velocity–time graphs

If we measure velocity and time we can draw graphs. These graphs allow us to see more clearly the motion of an object changes with time. It also allows us to make calculations from the graph to give additional information.

Three types of motion are shown in Figure 5.9:

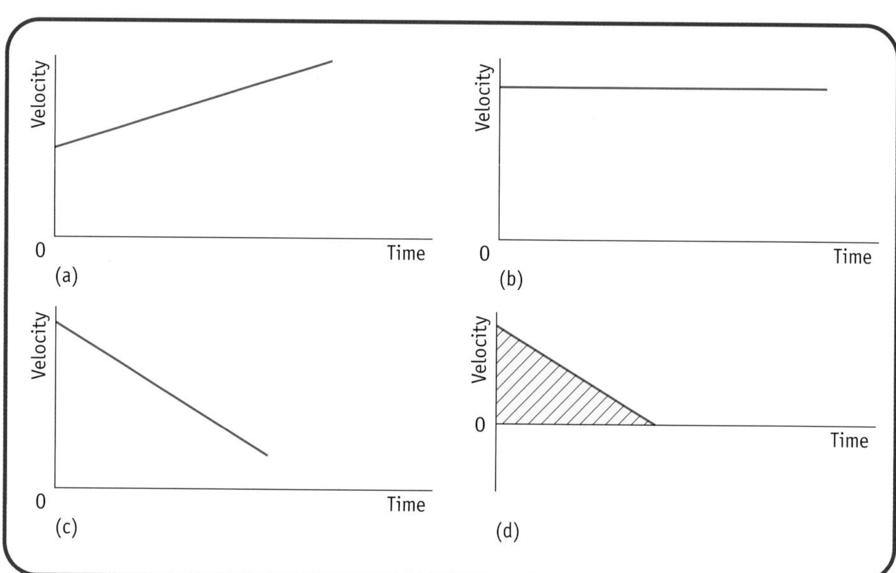

Figure 5.9 *a) Uniformly increasing velocity, b) Constant velocity, c) and d) Uniformly decreasing velocity*

a) *Uniformly increasing velocity.* This is a straight line at an angle to the horizontal. The gradient of the graph allows us to calculate the acceleration.
b) *Steady or constant velocity.* This is a straight line parallel to the time axis.
c) and d) *Uniformly decreasing velocity.* This is a straight line heading down towards the time axis. Again the gradient can be used to calculate the acceleration.

Key points

- The area under any part of a velocity–time graph is the displacement of the object. Normally the shape of the graph will consist of a rectangle and/or triangles (Figure 5.7d).
- If the object is travelling in one direction in a straight line then distance travelled = displacement.
- To calculate the average speed of an object when more than one type of motion is involved, draw a speed–time graph and find the area under it, i.e. distance travelled, then:

$$\text{average speed} = \frac{\text{distance travelled}}{\text{time taken}}$$

- For an object moving only with constant acceleration or constant deceleration:

$$\text{average velocity} = \frac{\text{initial velocity} + \text{final velocity}}{2}$$

Example

The velocity–time graph for a car travelling along a road is shown in Figure 5.10

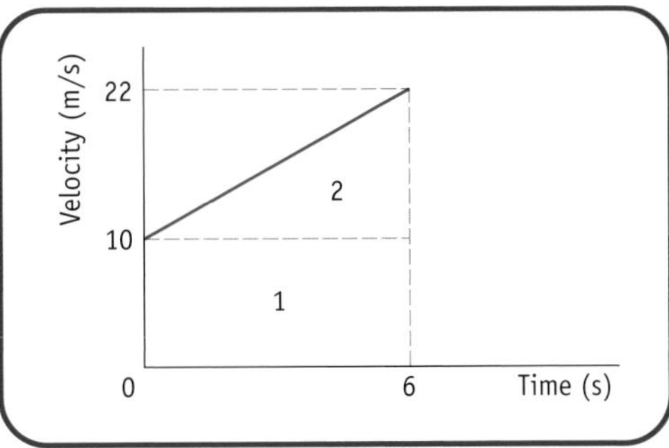

Figure 5.10 *Velocity–time graph for a car travelling along a road*

a) Calculate the acceleration of the car.
b) Calculate the distance travelled by the car.
c) Calculate the average speed of the car.

Solution

a) The acceleration is given as $a = \dfrac{v - u}{t}$

where $u = 10$ m/s
$v = 22$ m/s
and $t = 6$ s

$$a = \frac{22 - 10}{6} = \frac{12}{6} = 2 \text{ m/s}^2$$

b) Distance = area under velocity–time graph
= area 1 + area 2
$= 10 \times 6 + \dfrac{1}{2} \times 12 \times 6$
$= 60 + 36$
$= 96$ m

c) Average speed $= \dfrac{\text{distance}}{\text{time}} = \dfrac{96}{6} = 16$ m/s

Physics facts and key equations for kinematics

- average speed $= \dfrac{\text{distance}}{\text{time}}$.

  Unit of speed is m/s.
- Instantaneous speed is the same calculation as average speed but the time interval is very short.
- Vectors have magnitude and direction but scalars have only magnitude.
- Displacement and velocity are vectors since they have direction as well as magnitude but distance and speed are scalars.
- Acceleration $a = \dfrac{v - u}{t}$ and has the unit of m/s^2.
- Displacement is the area under a velocity–time graph.
- Acceleration is the gradient of a velocity–time graph.

Questions

1 Copy and complete the table below

Distance (m)	Time (s)	Average Speed (m/s)
24	6.0	
33	5.5	
	7.0	17
	8.5	11
160		8
420		14

2 A car travels a distance of 35 m in a time of 5 s. Calculate its average speed.

3 A cyclist has an average speed of 8.5 m/s and travels for 7 s. How far has the cyclist travelled in this time?

4 A car travels 46 000 m in one hour. What is the average speed of the car in m/s?

5 A vehicle is travelling at a constant speed of 23 m/s. It travels a distance of 345m. Calculate the time taken by the vehicle to travel 345 m.

6 A car travels at a constant speed of 15 m/s for 5 minutes. Calculate the distance travelled in this time.

7 In a police speed check, photographs are taken 0.5 s apart. The marks are 2 m apart and the car crosses 4 marks. The speed limit for this section of the road is 13.3 m/s.

a) Show that the car was breaking the speed limit.

b) Is this an average or instantaneous speed?

8 During a charity walk, some students walk round a circular track of circumference 30 m. A student takes 15 s to complete a circuit. Calculate the average speed and the average velocity of the student.

9 a) Explain the difference between vectors and scalars and give an example of each.

b) A car travels 40 m east then 30 m south. The total distance travelled is 70 m but the displacement is 50 m at 127°. Explain the difference between distance travelled and displacement.

c) A swimming pool is 25 m long. A swimmer swims from one end of the pool to the other and then turns and swims back to the start.

(i) What distance has she travelled?

(ii) What is her displacement from the start?

d) An ocean liner is travelling east at 12 m/s. A passenger runs south at 4 m/s. Calculate the resultant velocity of the passenger.

10 On a walk home from college you travel 200 m north, then 450 m east and finally 620 m south. What is your final displacement from the college?

11 An aircraft is flying east at 125 m/s. The wind is blowing from the north at 25 m/s. Calculate the resultant velocity of the aircraft relative to the ground.

12 A rowing boat is being rowed south on a river at a speed of 12 m/s. The river is flowing east at a speed of 7 m/s. Calculate the resultant velocity of the rowing boat relative to the bank of the river.

13 Use the table on page to calculate the acceleration of the following cars a) Honda Civic b) Jaguar c) Vauxhall Astra d) One of your choice

14 A new car is advertised as 'goes from 0 to 60 mph in 15 s'

a) What does this information tell you about the car?

b) How does the acceleration of this car compare with those answers to Q. 13 a), b) and c)?

15 A cyclist is travelling along a straight road and changes speed from 3m/s to 5m/s in 12 s. Calculate his acceleration.

16 Copy and complete the table below.

Initial Velocity (m/s)	Final velocity (m/s)	Time (s)	Acceleration (m/s^2)
0	25	10	
20	30	5	
25	45		2
30	14	8	

17 A car is travelling along a straight road. The speed of the car changes from 12 m/s to 30 m/s in 15 s. Calculate the acceleration of the car.

18 A car on the straight part of the motorway is being driven at 30 m/s. The driver sees a hazard ahead. She slows down to 25 m/s in a time of 4 s. Calculate the car's deceleration.

19 The graph shown in Figure 5.11 shows how the velocity of a skateboarder varies with time.

a) Describe the motion of the skateboarder for the two parts of the journey.
b) Calculate the acceleration of the skateboarder.
c) Calculate the distance travelled by the skateboarder after 15 s.

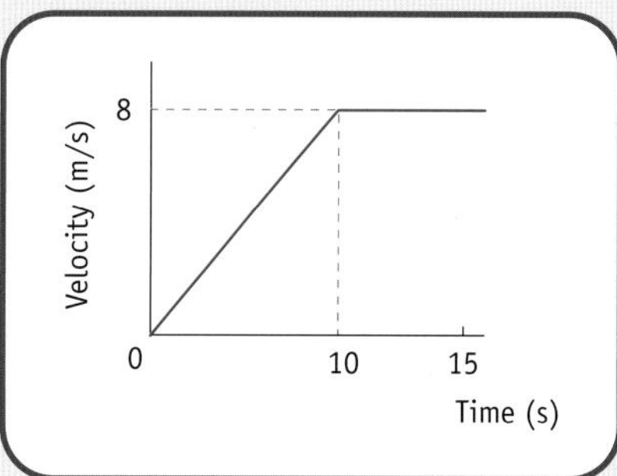

Figure 5.11

20 The velocity time graph shown in Figure 5.12 is for a lorry travelling along a straight road.

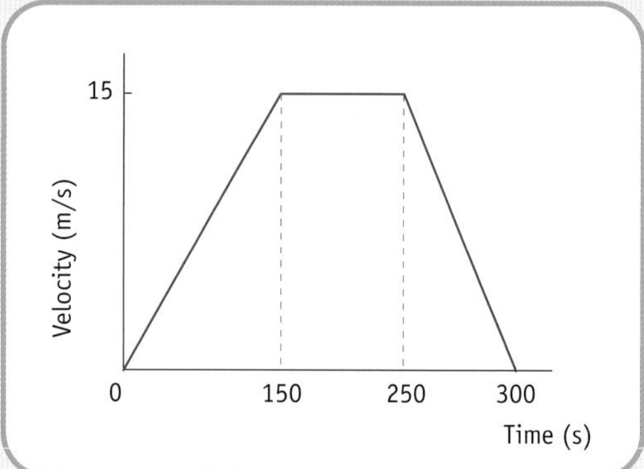

Figure 5.12

a) Describe the motion of the lorry over the three stages of the journey.
b) Calculate the deceleration of the lorry.
c) Calculate the distance travelled after 300 s.

21 A cyclist starts from rest and travels along a straight road for 10 s to reach a velocity of 25 m/s. He then travels for a further 8 s at this speed.

a) Sketch a graph showing how his velocity varies with time for the journey. Numerical values are required on both axes.
b) Calculate his acceleration.
c) Calculate the distance travelled after 18 s.

22 The graph shown in Figure 5.13 shows the motion of a car during a journey. Describe the motion of the car during the different stages.

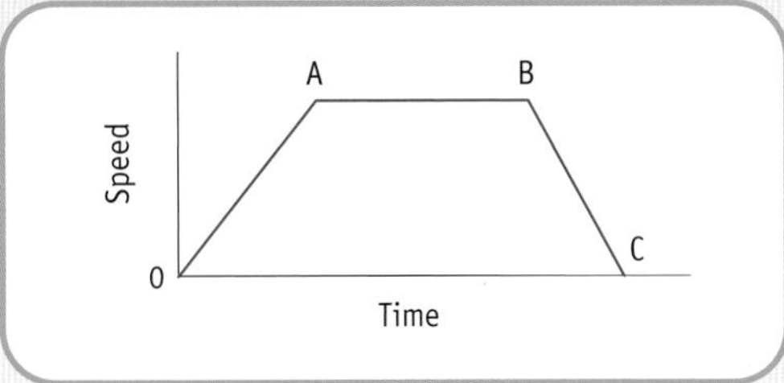

Figure 5.13

23 A car is travelling along a straight road at 10 m/s and increases its speed to 19 m/s in 3 s. Calculate the acceleration of the car.

24 A lorry brakes uniformly from 20 m/s to 8 m/s in 2.4 s. Calculate the deceleration of the lorry.

25 The velocity-time graph for a car journey is shown in Figure 5.14.

a) Calculate the acceleration of the car.

b) Calculate the car's displacement.

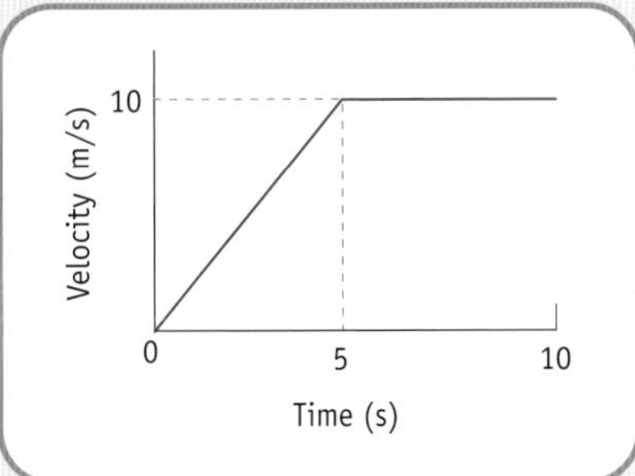

Figure 5.14

26 An electric car travels slowly along a straight road. The velocity-time graph for the car is shown in Figure 5.15. Calculate the acceleration of the car.

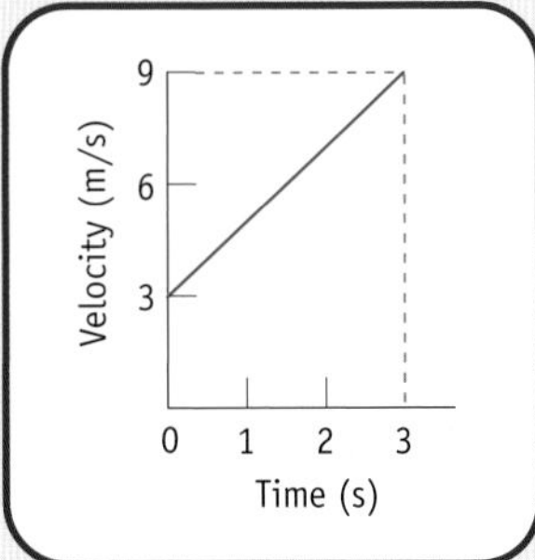

Figure 5.15

6 Forces

At the end of this chapter you should be able to...

1 Describe the effects of forces in terms of their ability to change the shape, speed and direction of travel of an object.
2 Describe the use of a newton balance to measure force.
3 State that weight is a force and is the Earth's pull on the object.
4 Distinguish between mass and weight.
5 State that weight per unit mass is called the gravitational field strength.
6 Carry out calculations involving the relationship between mass, weight and gravitational field strength including situations where the value of the gravitational field strength is not 10 N/kg.
7 State that the force of friction can oppose the motion of a body.
8 Describe and explain situations in which attempts are made to increase or decrease the force of friction.
9 State that force is a vector quantity.
10 State that forces which are equal in size but act in opposite directions on an object are called balanced forces and are equivalent to no force at all.
11 Explain the movement of objects in terms of Newton's first law.
12 Describe the qualitative effects of change of mass or of force on the acceleration of an object.
13 Define the newton.
14 Use free body diagrams to analyse the forces on an object.
15 State what is meant by the resultant of a number of forces.
16 Use a scale diagram or otherwise to find the resultant of two forces acting at right angles to each other.
17 Carry out calculations using the relationship between acceleration, resultant force and mass and involving more than one force but in one dimension only.
18 Explain the equivalence of acceleration due to gravity and gravitational field strength.
19 Explain the curved path of a projectile in terms of the force of gravity.
20 Explain how projectile motion can be treated as two independent motions.
21 Solve numerical problems on projectiles.

Forces and motion

The effects of forces

When an object is pushed or pulled, a force is exerted on it.

Forces can have three effects on an object:

- A force can change the speed of a moving object.
- A force can change the direction of a moving object.
- A force can change the shape of (deform) an object.

These effects will depend on the size of the force applied to the object.

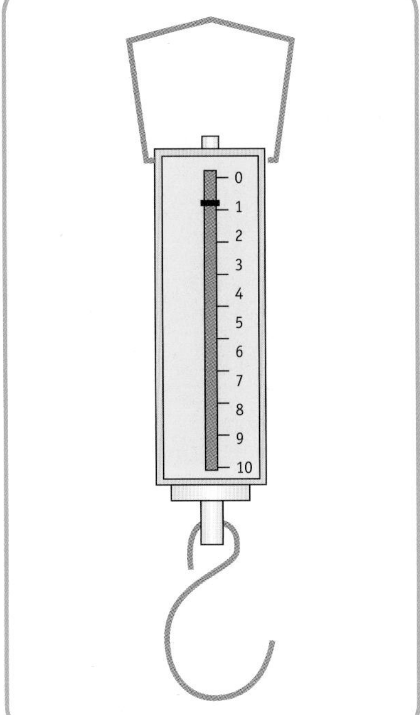

Figure 6.1 *The Newton balance*

Measuring force

Springs can be used to measure force:

1 A spring stretches evenly – each time the same force is applied to the spring, the spring stretches by the same amount.
2 A spring returns to its original length when the force is removed. If the force is too great the spring will not return to its original length when the force is removed.

We measure forces using a Newton balance. The unit of force is the **newton** (N) (Figure 6.1).

Frictional forces

Moving vehicles such as cars can slow down due to forces acting on them. These forces can be due to the road surface, the tyres, the brakes, or even air resistance. The force that tries to oppose motion is called the force of friction. A frictional force always acts when particles are sliding across one another and will oppose any motion. It will try to slow down an object like a car regardless of the direction in which the car is moving along a road.

Did you know?

Racing cars do not have treads on their tyres because they normally only race on dry days. There is no need for tread to be cut in the tyres to direct water away from the road surface. The greater amount of rubber allows better grip on the surface and better road handling (Figure 6.2).

Figure 6.2 *Treads on a racing car's tyres*

Car brakes

Car brakes operate by slowing down the car. When the brake pedal is pressed it causes the expander to push the brake shoes against the drum. If the brakes lock it will be very difficult to steer (Figure 6.3).

Figure 6.3 *Braking system*

Many cars are fitted with an anti-lock braking system called ABS (see Figure 6.4). This uses sensors located at each wheel to detect when it is about to lock during braking. The brakes then are rapidly released and then reapplied.

Reducing friction

Lubrication

The frictional force between two surfaces moving against each other can be reduced by lubricating the surfaces. This generally means that oil can be placed in between two metal surfaces. This happens in car engines and reduces wear on the engine since the metal parts are not actually meeting each other but have a thin layer of oil between them.

Streamlining

Modern cars are designed to offer as little resistance (drag) to the air as possible. This is the friction of the air on the car. The force depends on both the speed of the car and the frontal area presented to the air flow. To reduce this friction the designers try to streamline the vehicle in a variety of ways. This streamlining is measured by a number called the drag coefficient, C_d. The larger the C_d number, the greater the resistance to air flow. The calculation of C_d is difficult and involves the vehicle being placed in a wind tunnel and smoke flowing over it. The more turbulence that is created the higher the value of C_d.

Most C_d values range from 0.3 to 0.4 since cars have similar shapes. The drag coefficient can be reduced in a number of ways:

- Reducing the front area of the car.
- Using door mirrors instead of wing mirrors.

Figure 6.4 *Anti-lock braking system*

- Having a smooth round body shape.
- Using aerials made as part of the car windows.

In addition you should not carry a roof rack unless it is needed, since this increases the frontal area.

Many modern cars look very similar since they are all designed to this brief. Older cars often have more distinctive features but they are less fuel efficient.

Newton's first law

Sir Isaac Newton was appointed Lucasian Professor of Mathematics at Cambridge, where he worked on mathematics and astronomy. The physicist Stephen Hawking, author of *A Brief History of Time*, holds the same post today (Figure 6.5).

Figure 6.5 *Professor Stephen Hawking.*

Newton proposed three laws of motion. His first law states:

An object will remain at rest or move at constant speed in a straight line unless acted on by an unbalanced force.

Using this law we can explain the following situations:

1 An ocean liner travelling at a steady speed through the water (Figure 6.6). The resistive force offered by the water will balance out the force exerted by the engine.
2 A motorcycle moving at a constant speed along a level road and no matter how hard the driver tries to accelerate the motorcycle will not increase its speed (Figure 6.7). Again the resistive forces produced by the air resistance will balance the force produced by the engine.
3 A low friction vehicle on an air track will keep moving in a straight line at constant speed unless acted on by a force. This force could be the end stop of the track or a person touching the vehicle. This is because there are no forces acting along the air track.

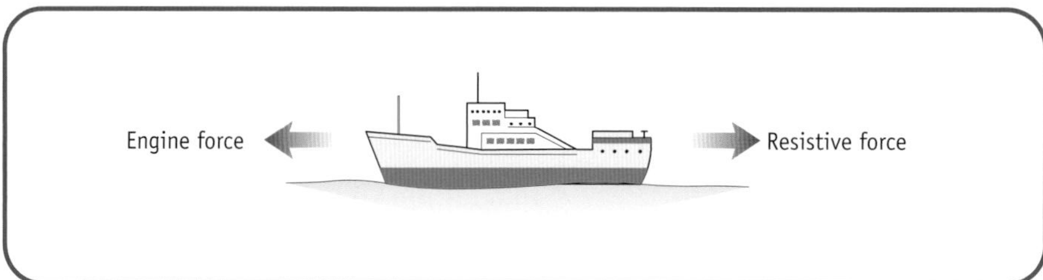

Figure 6.6 *Forces acting on an ocean liner*

Figure 6.7 *Forces acting on a motorcycle*

Mass and weight

Mass

Mass is the quantity of matter forming an object. This depends on the number and type of atoms making it up. The mass of an object remains constant. If you go to different parts of the Earth or even to different planets you have the same mass since there are no changes to the number and type of atoms in your body.

Force of gravity

Force of gravity is the downward pull of the earth on an object:

force of gravity = pull of Earth on an object

= gravitational force on an object

= weight of an object

Weight

The weight of an object depends on:

a) Mass.

b) Where you are in the solar system.

It is a force and is measured in newtons (N).

$$\frac{\text{weight of an object}}{\text{mass of an object}} = \frac{W}{m} = \text{constant}$$

This constant is called the gravitational field strength (g).

Gravitational field strength varies depending on where you are in space. For instance, on Earth $g = 10$ N/kg but on the Moon $g = 1.6$ N/kg and on Jupiter $g = 26.4$ N/kg. The unit of g is N/kg. Thus:

weight = mass × gravitational field strength

$W = mg$

Example

What is the weight of a 75 kg person on Earth?

Solution

$W = mg$

$= 75 \times 10$

$= 750$ N

Forces and Supported Objects

We know from Newton's first law that if an object either remains stationary, or moves upwards or downwards at constant speed, then the upward force is equal in size to the weight of the object but acts in the opposite direction. This is known as balanced forces.

Example 1

A stationary object of mass *m* hangs from a rope. The weight of the object is mg (N) and acts downwards, but this is counterbalanced by a force of the same size acting upwards due to the tension (T) in the string (Figure 6.8).

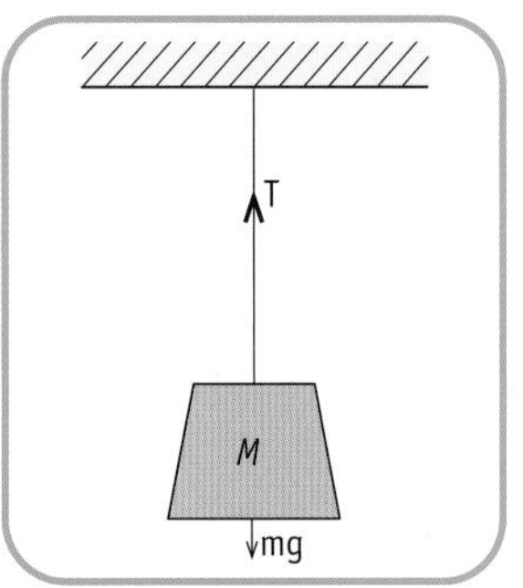

Figure 6.8 *A suspended object*

Example 2

A book rests on a shelf. The shelf gives an upward force, called the normal reaction (R), which is equal in size to the weight (mg), and therefore the forces are balanced (Figure 6.9).

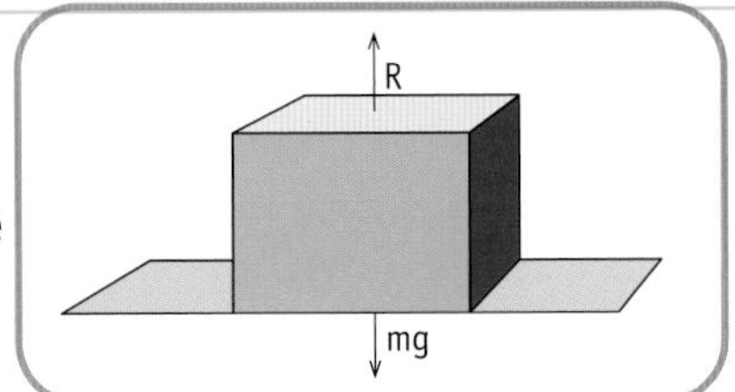

Figure 6.9 *A book resting on a shelf*

We can also see an example of balanced forces in action in a different situation:

Aircraft

The engines provide a forward force or **thrust** which accelerates the aircraft forward. However, as it moves faster the air resistance, or drag, increases until the forces in a straight line (horizontally) are balanced. The aircraft would move at constant speed (Newton's first law). When in level flight at constant speed the **lift** will balance the weight (Figure 6.10). Horizontally the thrust is equal in size to the drag force. This produces balanced forces.

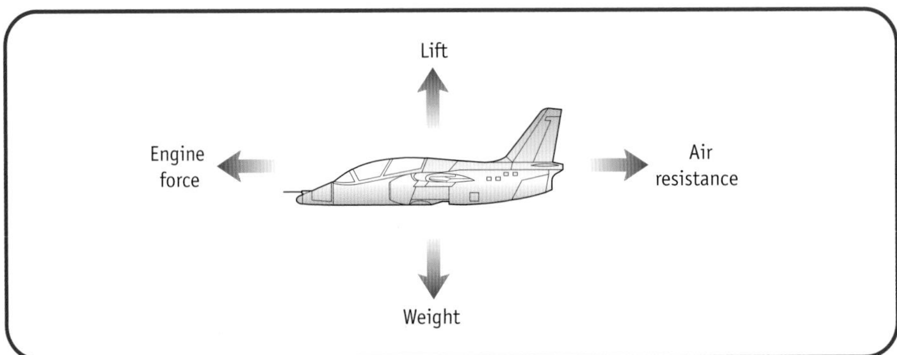

Figure 6.10 *Forces acting on an aircraft*

Inertia

All objects are unwilling (or reluctant) to change their motion. This means that an unbalanced force is required to change the motion of an object.

This reluctance of an object to change its motion is called its inertial mass or inertia. It depends on mass. The larger the mass, the larger the inertia and the more unwilling the object is to change its motion. It is easier to stop a lighter object than a more massive one travelling at the same speed.

Seat belts

During the sudden braking of a car, any unrestrained object will continue to move in a straight line at the car's original speed. This happens due to Newton's first law. Any objects will probably collide with some part of the interior, causing damage or injury.

A seat belt applies a force in the opposite direction of motion causing rapid deceleration of the wearer. The webbing straps are designed to have a certain amount of 'give' so that the sudden force applied to the person does not cause injury (Figure 6.11).

Figure 6.11 *Seat belt*

Adult seat belts are not suitable for young children since the young children tend to slip down in the seat. This effect is called submarining and can be prevented by using a special design for children.

Air bags produce a similar effect and the large area of the bag prevents chest injuries due to the steering wheel. This is because the larger area of the bag spreads out the force, reducing the pressure.

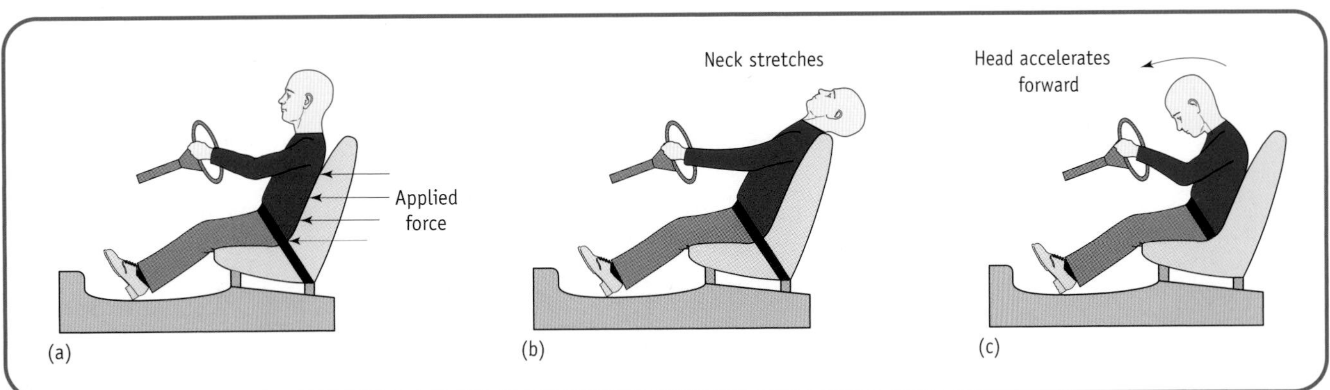

Figure 6.12 *Whiplash injury*

Whiplash injuries occur when the head of the person in an accident stays stationary but the trunk moves forward. This is shown in Figure 6.12. In (a) the trunk is moved forward since the vehicle is struck from behind. The inertia of the head (b) causes it to stay in place while the trunk of the body moves forward. This causes stretching in the neck region. As the vehicle slows the head is accelerated forward. In a typical collision, at 15 m/s (about 34 mph) the collision with the vehicle will cause these injuries to occur in milliseconds.

Did you know?

Why are headrests needed in cars?

Headrests are in cars for the same reason as seatbelts, that is, safety. They prevent the head from going backwards when the car is struck from the rear. When the collision occurs the vehicle and its seat exert a force on the person's body accelerating it forward. The seat does not push the head forward so the head will keep its original motion relative to the ground and will appear to move back. The headrest pushes the head forward with the rest of the body to prevent serious injuries to the back and spinal system.

Balanced and unbalanced forces

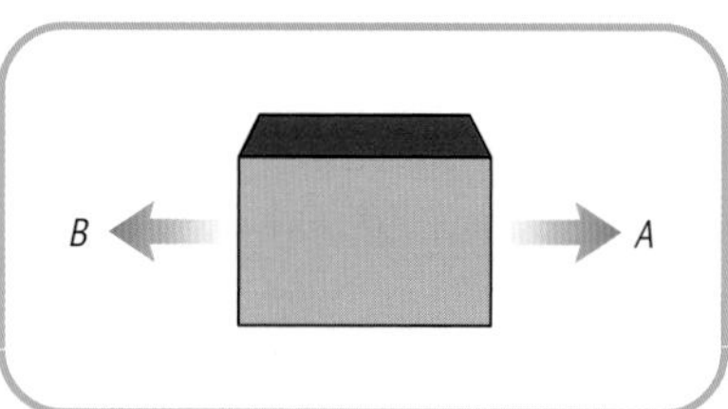

Figure 6.13 *Balanced forces acting on an object*

Balanced forces acting on an object are equal in size but act in opposite in directions.

They cancel each other out and thus are the equivalent of zero force acting on the object (Figure 6.13).

An unbalanced force acting on an object causes it to speed up or slow down.

When A and B are both same size, the same size of force is applied to each side of the object. The forces are balanced. If the object is at rest it will stay at rest.

If the object is moving at a constant speed in a straight line it will continue at that speed in a straight line.

When force A is greater than force B, the object will accelerate to the right.

Force, mass and acceleration

A mass m is accelerated by an unbalanced force (F_{un}) . If we double the unbalanced force we double the acceleration.

An unbalanced force(F_{un}) acts on an object of mass m and if you double the mass then you will halve the acceleration.

This is Newton's second law.

When an object is acted on by a constant unbalanced force, the object moves with constant acceleration in the direction of the unbalanced force:

$$\text{unbalanced force} = \text{mass} \times \text{acceleration}$$

In symbols $F_{un} = ma$

In units where F_{un} is in newtons (N), m is in kilograms (kg) and a is in metres per second squared (m/s^2).

Example

An unbalanced force of 75 N acts on a mass of 15 kg. Calculate the acceleration of the mass.

Solution

$F_{un} = 75$ N, $m = 15$ kg, $a = ?$

$F_{un} = ma$

$$a = \frac{F_{un}}{m} = \frac{75}{15} = 5 \text{ m/s}^2$$

Example

A sledge is being pulled by an unbalanced force of 100 N. The sledge accelerates at 5 m/s^2. What is the mass of the sledge?

Solution

$F_{un} = 100$ N

$a = 5$ m/s^2

$F_{un} = ma$

$100 = m \times 5$

$$m = \frac{100}{5}$$

$= 20$ kg

Example

The force exerted by the engine of a car is 2000 N. The car has a mass of 1000 kg. There is a frictional force of 100 N acting on the car. Calculate the acceleration of the car (Figure 6.14).

Figure 6.14

Solution

The unbalanced force acting on the car $F_{un} = 2000 - 100 = 1900$ N

$$a = \frac{F_{un}}{m} = \frac{1900}{1000} = 1.9 \text{ m/s}^2$$

Acceleration due to gravity

The force of gravity causes all objects to accelerate as they fall back to earth. This is due to the earth being massive and it exerts a very strong force on these small objects. This gravitational force is the one force that we cannot switch off.

Consider a mass m which is released from a point near the surface of the Earth. We will assume that the effect of air resistance is negligible. The only force acting on the object is its weight. This is an unbalanced force and so the object will accelerate towards the surface of the Earth.

Unbalanced force force on the object = mg = ma

With mass being constant a = g

Acceleration due to gravity = gravitational field strength

$$\text{gravitational field strength} = \frac{\text{force of gravity on an object}}{\text{mass of object}}$$

$$g = \frac{\text{weight}}{\text{mass}} = \frac{W}{m} = \text{acceleration due to gravity}$$

Frictional forces in a fluid

A fluid is a liquid or gas. In a fluid the frictional force on an object travelling through it increases as the speed of the object increases. If two spheres are dropped into different liquids such as oil and water the one in the water will reach the bottom first.

In both cases the spheres accelerate and then move with constant speed (terminal velocity). When the sphere is accelerating there is an unbalanced force acting on it. When the terminal velocity is reached the forces are balanced. It is reached sooner in the thick liquid than in water. This shows that the frictional force in a liquid depends on the type of liquid.

The motion of an object falling through any fluid can be divided into three parts:

1 Initially an unbalanced force acts on the object due to its weight and the object falls with a constant acceleration of 10 m/s^2, which is the acceleration due to gravity.
2 After a short time the frictional force begins to act and alters the motion. This force will be increasing as the speed of the object increases. There is a smaller unbalanced force ($F_{un} = W - F_r$) and the acceleration is therefore less than 10 m/s^2. This acceleration will continue to decrease as the frictional force increases.
3 Finally the frictional force balances the weight of the object. We now have balanced forces. The object now falls at a constant speed in a straight line. The object has reached its greatest speed. This is called its terminal speed or terminal velocity (Figure 6.15).

A free-falling parachutist is a typical example of such forces in action. Figure 6.16 shows a graph of speed against time for a parachutist falling in free fall out of an aircraft and then opening the rip cord some time later. Each part of the graph is explained opposite.

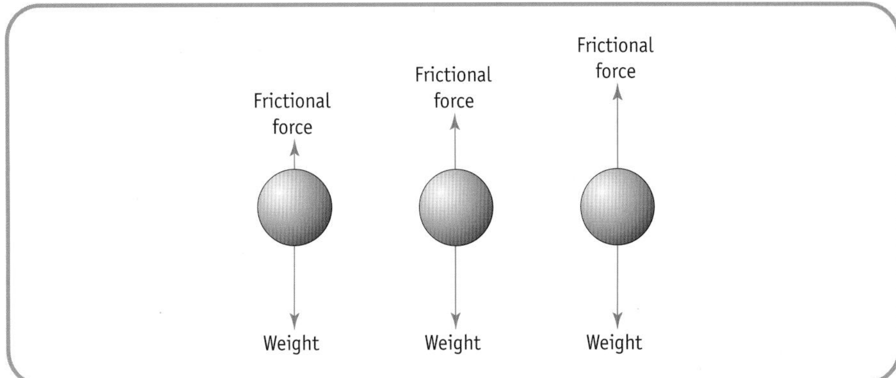

Figure 6.15 *Terminal velocity*

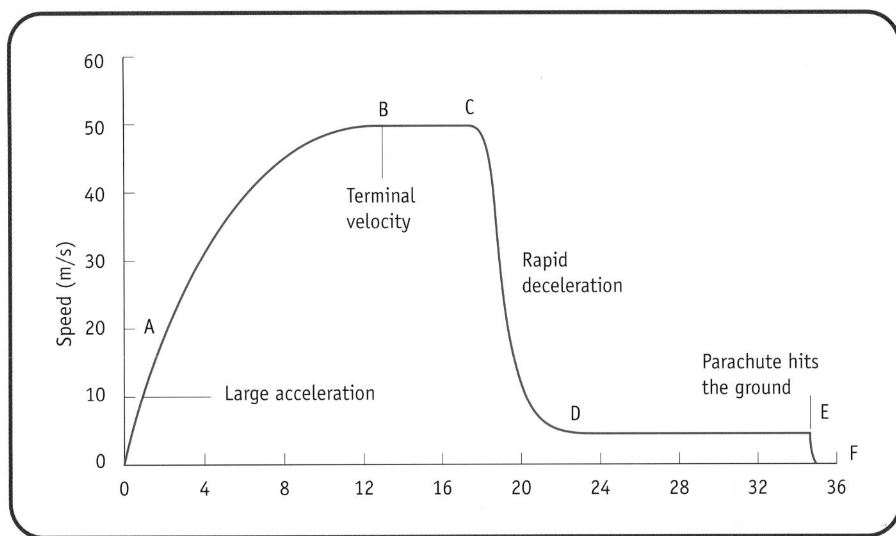

Figure 6.16 *Parachutist's descent*

OA: constant acceleration due to gravity

AB: decreasing acceleration as frictional force acts:

unbalanced force = weight − frictional force

BC: constant speed as frictional force upwards = weight downwards

CD: non-uniform deceleration due to parachute opening and increasing frictional force

DE: constant speed as frictional force upwards = weight downwards

EF: parachutist hits the ground.

Did you know?

Free Fall

Free fall only truly occurs when there is no air resistance. If the air resistance is very small then after 10 s your speed would be 100 m/s and you would have fallen almost 500 m. In a free fall parachutist then the maximum speed you can reach is about 150 to 200 km/h or between 42 m/s and 56 m/s. The final terminal velocity will depend on the person's size or how aerodynamic they are!

Force as a vector

When two forces act on an object the combined effect depends on the size and direction of the forces.

Force is a vector; that is, it has magnitude and direction.

The combined effect of the forces is called the resultant force. The resultant force can be found using the following technique:

- Sketch a diagram of the situation.
- Draw known lines to represent the forces and the angles.
- The lines are shown with an arrow to indicate direction.
- The tail of one vector should join the head of another vector.
- The resultant force is represented by a line drawn from the tail of the first vector to the head of the last vector.
- Use a suitable scale or use Pythagoras.
- Calculate the resultant force.
- Calculate or measure the angle of the resultant force.

Example

A parachutist is falling to the ground. She has a weight of 60 N.The wind force on the parachutist is 15 N westwards. Calculate the resultant force on the parachutist.

Solution

Using the diagram in Figure 6.17, lines are drawn to represent the two forces. The resultant force is marked. Since the two forces are at right angles the magnitude of the force can be calculated using Pythagoras. The angle of the resultant force can be found by measurement or calculation.
The resultant force is 61.8 N at 194°.

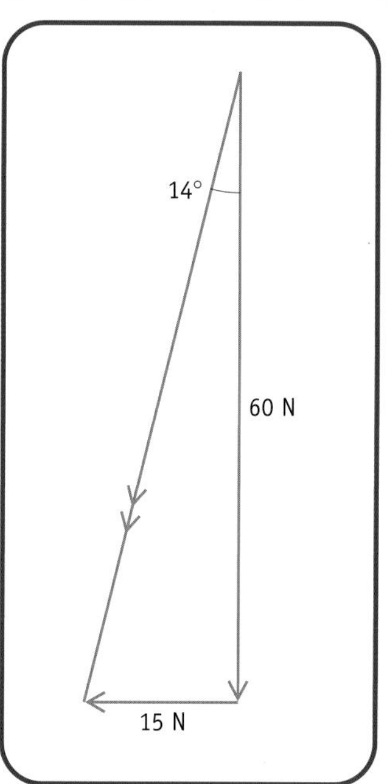

Figure 6.17

Acceleration due to gravity

An object falling near the surface of the Earth accelerates (if the effects of air resistance are negligible). This acceleration is called the acceleration due to gravity (g).

Consider two objects of mass 5 kg and 10 kg falling near the surface of the earth. We are assuming that there is no friction present.

Weight = force of gravity

$= mg$

$= 5 \times 10$

$= 50$ N

weight = force of gravity

$= mg$

$= 10 \times 10$

$= 100$ N

Since we are assuming no friction is present:

unbalanced force $= F_{un}$ = weight

$= 50$ N

$$a = \frac{F_{un}}{m} = \frac{50}{5}$$

$= 10$ m/s^2

unbalanced force $= F_{un}$ = weight

$= 100$ N

$$a = \frac{F_{un}}{m} = \frac{100}{10}$$

$= 10$ m/s^2

This shows clearly that the acceleration due to gravity in the absence of friction (air resistance) is the same for *all* objects, no matter what their mass.

The acceleration due to gravity on a planet has the same number value as the gravitational field strength on the planet. For example:

On a planet

acceleration due to gravity = gravitational field strength

On Jupiter gravitational field strength = 25 N/kg

acceleration due to gravity = 25 m/s^2

Measuring the acceleration due to gravity

Using a light gate, computer and a piece of card cut as a mask we can calculate the acceleration due to gravity. The card is dropped through the light gate, from different heights, and the acceleration is measured in each case. It is found that the height the object is released from does not affect the acceleration.

For the Earth, the acceleration due to gravity $g = 10$ m/s^2.

Weight on Earth and other planets

The weight of an object near the surface of a planet such as the Earth is the pull of the planet on the object. Like all forces, the weight is measured in newtons.

The pull of gravity on a falling object (the weight) can be calculated using the equation

$$W = mg$$

In words, weight = mass × gravitational field strength

g on earth = 10 N/kg

The pull of gravity per kilogram mass on a planet is called its gravitational field strength. Gravitational field strength is measured in newtons per kilogram (N/kg).

Projectile motion

There are a wide variety of objects in orbit around the Earth, ranging in size from satellites the size of a house to an astronaut's glove. To understand how it is possible for these objects to stay in orbit, we study objects moving horizontally and vertically at the same time. These are called projectiles.

If an object is projected horizontally at 16 m/s then its inertia (unwillingness to change its motion) should tend to keep it moving at 16 m/s horizontally (in a straight line). Due to the force of gravity (the object's weight) this is not possible and the object experiences a force pulling it downwards while it tries to move horizontally at constant speed. To overcome this conflict between its inertia and its weight, the object follows a curved path (Figure 6.18). This is called projectile motion. The motion of the projectile is therefore made up of two separate motions:

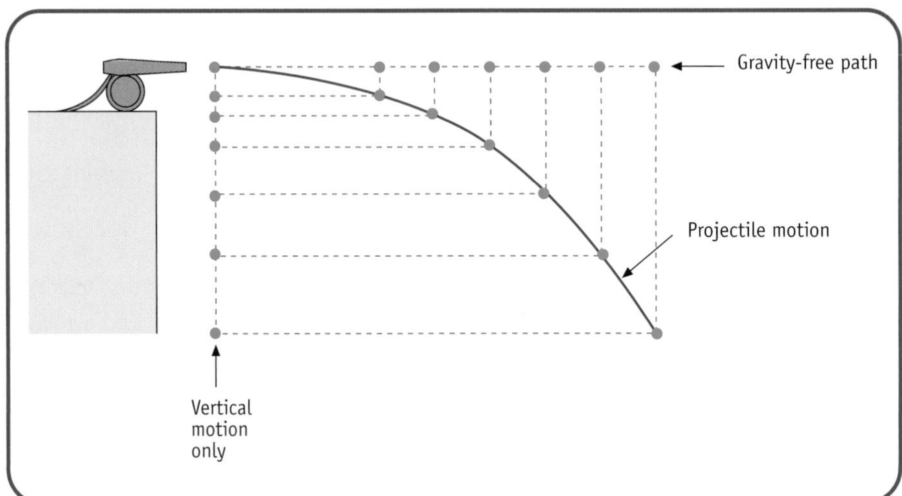

Figure 6.18 *Projectile motion*

1 A vertical motion under the influence of the force of gravity.
2 A horizontal motion under the influence of its inertia.

A special photograph is taken of a ball being projected, and a ball being dropped vertically at the same time, as shown in Figure 6.19.

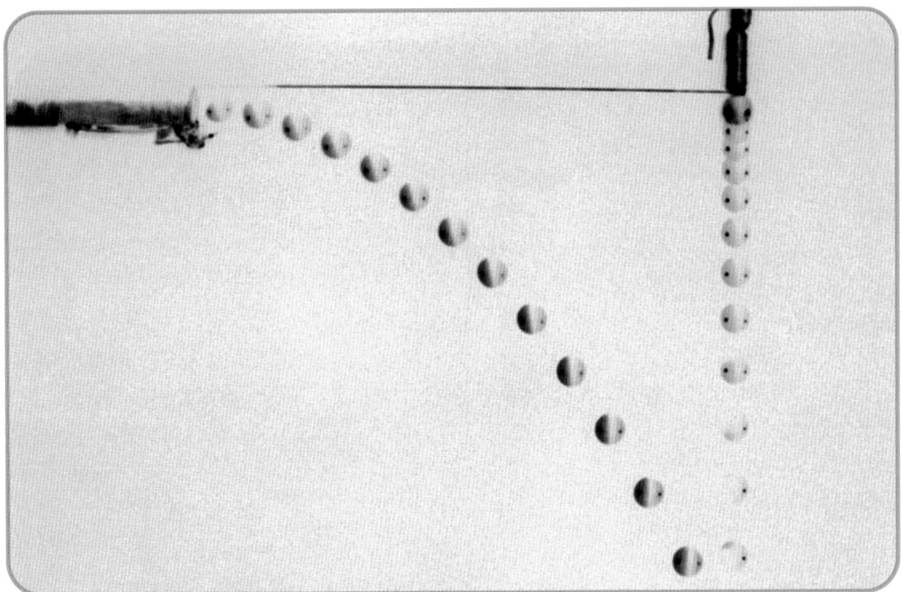

Figure 6.19 *Photograph showing a ball being projected, and a ball being dropped vertically*

The vertical motion of the projectile and the free-falling object are in step all the way down. This means that the vertical motion of the projectile is the same as a free-falling object; that is, both are accelerating downwards with a constant acceleration of 10 m/s^2. This is the acceleration due to gravity, which was discussed earlier.

The horizontal spacings between the images of the projectile are all equal. This means that the horizontal motion of the projectile is constant speed and is equal to the speed of projection.

The motion of a projectile can be treated as two independent motions:
- Constant speed in the horizontal direction.
- Constant acceleration in the vertical direction due to the force of gravity.

The speed–time graphs for the two motions are shown in Figure 6.20.

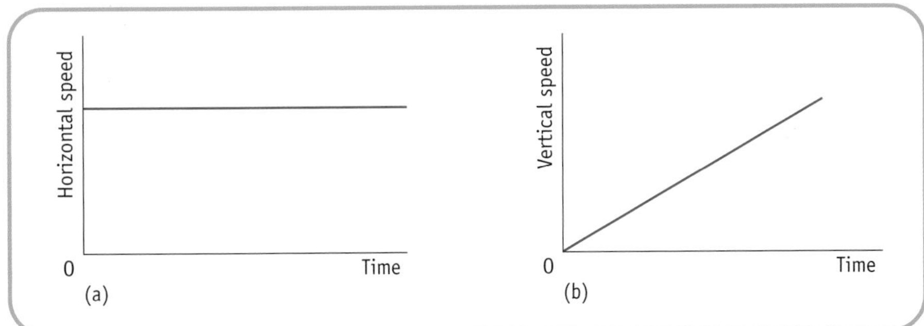

Figure 6.20 *a) constant speed in horizontal direction and b) constant acceleration in vertical direction*

When solving problems, we can treat the motion in the horizontal direction *separately* from the motion in the vertical direction.

Example

A flare is fired horizontally out to sea from a cliff-top at a speed of 40 m/s. The flare takes 4 s to reach the sea.

a) What is the horizontal speed of the flare after 4 s?

b) Calculate the vertical speed of the flare after 4 s.

c) Draw a graph to show how the vertical speed of the flare varies with time.

d) Use this graph to calculate the height of the cliff top above the sea.

Solution

a) Horizontal speed remains at 40 m/s.

b) Using $v - u = at$ for the vertical motion, where $u = 0$,

$$a = g = 10 \text{m/s}^2$$
$$t = 4 \text{ s}$$
$$v - 0 = 4 \times 10$$
$$= 40 \text{ m/s}$$

c) The graph is shown in Figure 6.21 opposite.

d) The height of the cliff is the area under the graph:

$$\text{distance} = \frac{1}{2} \times \text{base} \times \text{height}$$
$$= 0.5 \times 4 \times 40$$
$$= 80 \text{ m}$$

Height of the cliff above the sea is 80 m.

Newton's satellite

Newton, in considering the motion of the Moon around the Earth, carried out the following 'thought experiment'. Suppose a bullet is fired horizontally from a gun situated on top of a high mountain. The bullet will have two motions, which occur simultaneously:

a) A horizontal motion at uniform speed (if air resistance is negligible).

b) A vertical motion of uniform acceleration downwards under the action of the gravitational attraction between the bullet and the Earth.

As a result, the bullet will follow a curved path and will hit the ground no matter how fast it was fired, as long as the Earth is flat.

However, the approximation of the 'flat' Earth is only valid over a limited range. For a 'round' Earth it becomes important to take the curvature of the Earth into account for projectiles of long range (Figure 6.22).

If there was no force of gravity, the bullet would follow the path AB, because if no forces act on it, it must travel with a uniform speed in a straight line (Newton's first law).

Due to the gravitational field of the earth, the bullet falls continuously below this line. If the bullet falls below the line *AB* faster than the Earth's surface curves away under it, the bullet will still hit the Earth, e.g. at the point *C* or *D*, depending on the bullet's speed.

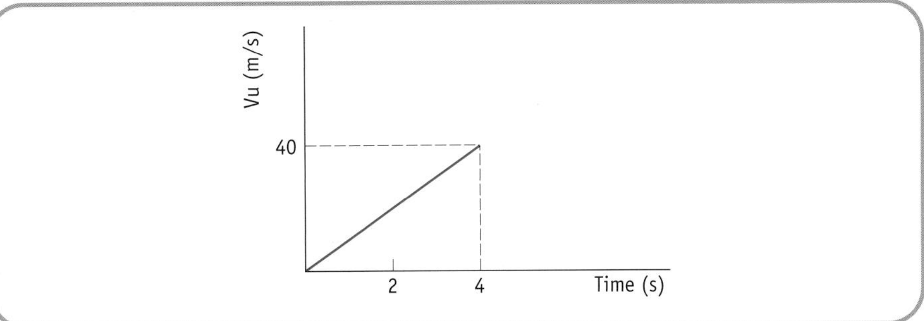

Figure 6.21 *Vertical speed of the flare*

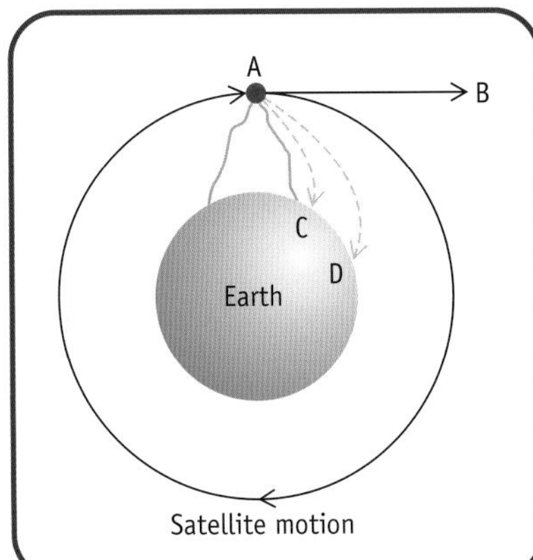

Figure 6.22 *Motion of a bullet fired from a gun from the top of a high mountain*

If the bullet was fired at such a speed that it fell vertically at the same rate as the Earth's surface curved away under it, then it would always be at the same height above the surface. The bullet would then never reach the surface, but would circle it at a constant altitude. This was the first theory of the artificial satellite.

Remember, Newton 'thought' about this experiment long before we had rockets or satellites. In fact the height of Newton's fictional mountain would here to be about ten times that of Mount Everest.

Did you know?

An aircraft has to drop food parcels at a specific point to help people in an emergency, for example during a famine or a flood. When should the parcel be dropped, before, over or past the target?
The best time is before the target since the parcel will still have the forward speed of the aircraft and will carry on at that forward speed while still accelerating downwards since it is a projectile.

Physics facts and key equations for forces

- Forces can change the shape, speed and direction of travel of an object.
- Mass is the amount of matter in a substance but weight is a force and is the Earth's pull on the object.
- Weight per unit mass is called the gravitational field strength.
- $W = mg$, where g is the gravitational field strength.
- The force of friction can oppose the motion of an object.
- Force is a vector.
- Equal forces acting in opposite directions on an object are called balanced forces and are equivalent to no force at all.
- Newton's first law states that an object will remain at rest or travel at constant velocity if the forces on the object are balanced.
- Newton's second law is $F_{un} = ma$.
- The resultant force is the one force which can replace all the forces acting on an object and have the same effect.
- Acceleration due to gravity = gravitational field strength.
- Projectile motion can be treated as two independent motions: horizontal motion – the object travels at constant velocity; vertical motion – the object has a constant downwards acceleration.

Questions

1 A dog weighs 75 N on Earth. What is its mass?

2 On the Moon, gravitational field strength is 1.6 N/kg. An astronaut has a mass of 100 kg on Earth. What is his mass on the Moon?

3 An astronaut has a mass of 95 kg on Earth. What will she weigh on the Moon?

4 A bowling ball has a weight of 70 N on Earth. What is its mass?

5 A family goes on holiday from Glasgow to Inverness and completes the journey at an average speed of 50 mph. When they repeat the journey with two light bicycles on the roof of the vehicle, the average speed is 45 mph.

a) Suggest a physics explanation for this effect.

b) What could be done to increase the average speed and still carry the bicycles?

6 The vehicles which were landed on the Moon were not streamlined unlike aircraft. Why was this not necessary with these vehicles? (Figure 6.23)

Figure 6.23

7 Copy and complete the table below.

Unbalanced Force (N)	Mass (kg)	Acceleration (m/s^2)
3000	250	
40 000	2000	
50		2
10 000		5
	35	7
	100	7.5

8 A sledge is sliding on ice which is practically frictionless. The force pulling the sledge is 130 N. The sledge accelerates at 3 m/s^2. What is the mass of the sledge?

9 A car accelerates along the road at 5 m/s^2.

a) The mass of the car is 1250 kg. What is the least engine force needed to produce this acceleration?

b) Why in practice will the engine force have to be greater that the value calculated in (a)?

10 A package is being pulled along a floor by a horizontal force of 60 N. The force of friction between the package and the floor is 12 N. The mass of the package is 12 kg.

a) Calculate the unbalanced force acting on the package.

b) What is the acceleration of the package?

11 A rocket of mass of 350 kg is accelerating vertically upwards.

a) What is the weight of the rocket?

b) The rocket is accelerating at 8 m/s^2. What is the size of the unbalanced force acting on the rocket?

c) What is the total force exerted by the rocket engines (the thrust)?

12 A ball is projected from the top of a cliff. The horizontal speed of the ball is 15 m/s. The ball takes 6 s to reach the ground.

a) Sketch graphs to show how:

(i) horizontal speed changes with time

(ii) vertical speed changes with time.

b) Calculate the horizontal distance travelled by the ball just as it hits the ground.

c) Use the graph in (a) (ii) or otherwise to calculate the vertical distance travelled.

13 A motorcyclist reaches a speed of 20 m/s. This is the maximum speed that can be reached.

a) Draw a diagram showing and naming the forces acting on the motorcycle at this time.

b) Are the forces balanced or unbalanced? Explain your answer.

14 A person has a mass of 60 kg. Use the data sheet to calculate the weight of this person standing:

a) on the surface of the Earth

b) on the surface of the Moon

c) on the surface of Mars.

15 A car of mass 800 kg is accelerating at 3.5 m/s^2. Calculate the unbalanced force on the car.

16 A car is being pushed on a cold day to start it moving. The car is moving at a constant speed. Name the force opposing the pushing force.

17 The force acting on an electric toy car due to the engine is 200 N. The frictional force acting against the car is 30 N. The mass of the car is 25 kg.

a) What is the value of the unbalanced force acting on the car?

b) Calculate the acceleration of the car.

18 A luggage trolley is being pulled by two ropes. Both ropes have forces of 75 N acting on them. If the two ropes are replaced by one rope which

has to have exactly the same effect, what is this replacement force called?

19 The acceleration due to gravity on a planet is 26 m/s^2. What is the gravitational field strength on this planet?

20 An aircraft is moving at 55 m/s horizontally when it drops supplies to help in a relief operation. The package takes 8 s to reach the ground.
a) Calculate how far the aircraft travels in 8 s
b) The parachute on the package fails to open. What is the vertical speed of the package just before it hits the ground?

Momentum and energy

At the end of this chapter you should be able to...

1 State Newton's third law.
2 Identify Newton pairs in situations involving several forces.
3 State that momentum is the product of mass and velocity.
4 State that momentum is a vector quantity.
5 State that the law of conservation of linear momentum can be applied to the interaction of two objects moving in one direction, in the absence of external forces.
6 Carry out calculations concerned with collisions in which all the objects move in the same direction and one object is initially at rest.
7 State that work done is a measure of the energy transferred.
8 Carry out calculations involving the relationship between work done, force and distance.
9 Carry out calculations involving the relationship between power, work done and time.
10 Carry out calculations involving the relationship between gravitational potential energy, mass, gravitational field strength and change in height.
11 Carry out calculations involving the relationship between kinetic energy, mass and velocity.
12 Carry out efficiency calculations involving output power, output energy and input power, input energy.

Newton's third law

When you try and strike a ball using a bat, the ball moves because a force is exerted on it. But you will also feel a force through the bat. Forces only exist in pairs which are equal in size but opposite in direction.

This is Newton's third law. It is sometimes written as:

Action and reaction are equal and opposite

Newton's third law involves two objects exerting forces on each other. In Figure 7.1 a parcel is placed on a table. The parcel weighs 500 N and this is the force that the parcel also exerts downwards on the table top. The table top exerts a force of 500 N upwards on the parcel.

Jet engines in aircraft and rocket propulsion use this principle. In both cases a high-speed stream of hot gases (produced by burning fuel) is pushed backwards from the vehicle with a large force (Figure 7.2) and a force of the same size pushes the vehicle forwards.

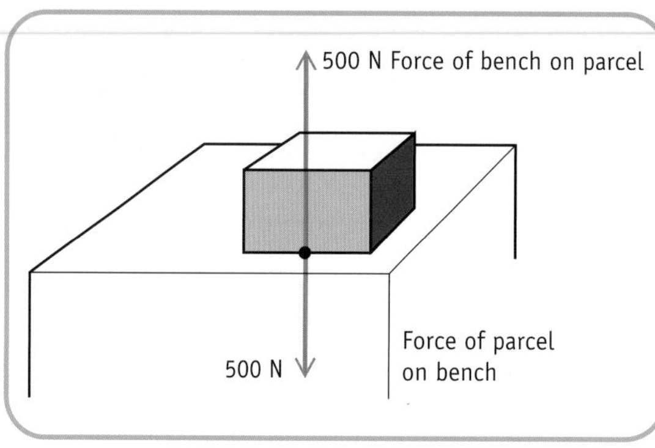

Figure 7.1 *Forces acting on a parcel resting on a table*

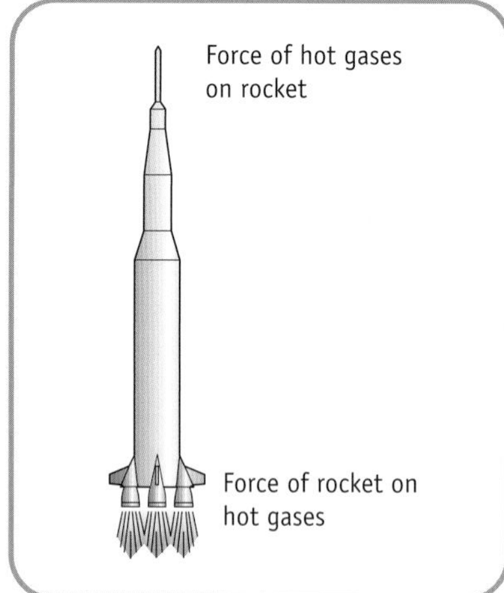

Figure 7.2 *Forces acting on a rocket*

Other examples:

- If you push against a wall, the wall pushes you backwards.
- In walking, you push your foot backwards but the floor pushes you forwards. The person and the floor of the Earth both apply equal amounts of force but because the Earth is so massive compared to your mass, you will accelelerate.
- In swimming, you push the water backwards but the water pushes you forward.

Put simply, this means that when *A* pushes *B*, *B* pushes *A* back with the same size of force.

Rockets and jet engines are able to produce motion.

Action: *vehicle* pushes *hot gases backwards* (downwards)

Reaction: *hot gases* push *vehicle forwards* (upwards)

At lift off for a rocket this force must be greater than the weight of the rocket. This will provide an unbalanced force upwards and the rocket will accelerate upwards.

Collisions and momentum

When a car collides with an object such as a stationary car the speed before the collision can sometimes be calculated from the tyre marks on the road. If two cars of different mass hit an identical stationary object then we would expect the heavier car to cause the greater damage. But what if the lighter car was moving faster?

To analyse these problems we use a concept called momentum:

momentum = mass x velocity.

unit = kg m/s

Momentum is a vector quantity.

The basic law of all collisions is that

total momentum before collision = **total** momentum after the collision

provided no external forces are acting.

This concept allows us to analyse simple collisions without knowing the forces that act during the collision.

To solve any problem in collisions

- Calculate the mass times the velocity of each part that is moving before the collision and add these parts together.
- Do the same for all the parts moving after.
- Finally use the law of total momentum being equal before and after collision.

To demonstrate momentum and collisions, an interface to a computer is used with two light gates. This set up can measure velocities before and after collisions (Figure 7.3).

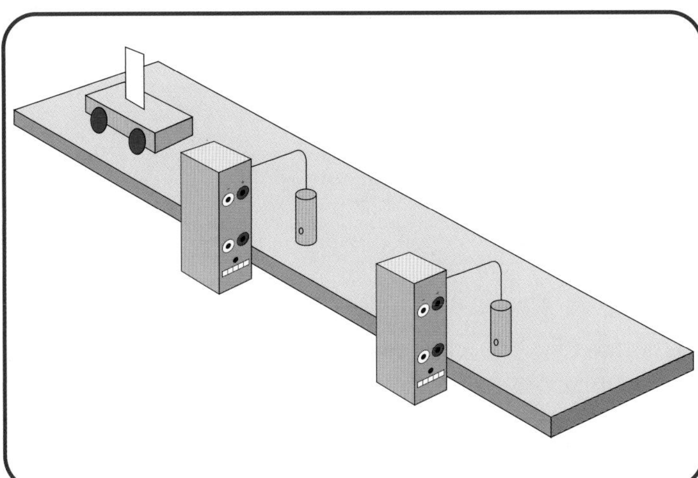

Figure 7.3a and b *Measuring initial and final velocity*

Example

A miniature wagon is moving at 6 m/s along a track. The mass of the wagon is 35 kg. The wagon collides and sticks to another wagon of mass 7 kg. What is the speed of the wagon after collision?

Solution

Before the collision, total momentum $= 35 \times 6 = 210$ kg m/s

After the collision, total momentum $= (35 + 7)\,v = 42\,v$

where *v* is the speed of the wagon after the collision.

Since total momentum before = total momentum after

$$210 = 42\,v$$

$$v = 5 \text{ m/s}$$

Example

A trolley of mass 10 kg is moving to the right at 6 m/s. It collides with another stationary trolley of mass 20 kg. After the collision both trolleys move off together. Calculate the speed with which they move off (Figure 7.4).

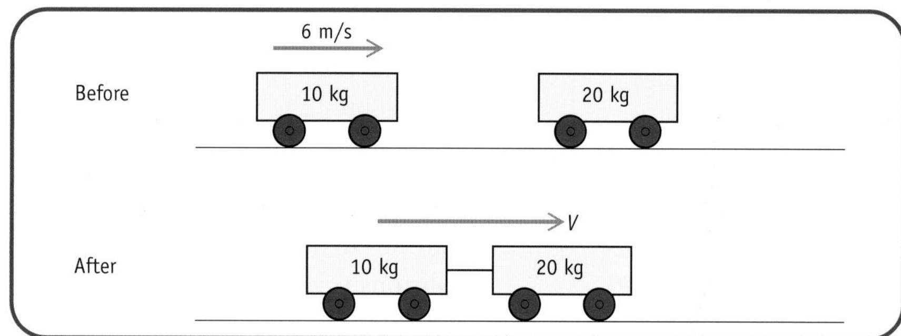

Figure 7.4

Solution

Before the collision, total momentum = 10 × 6 = 60 kg m/s
After the collision, total momentum = (10 + 20) v = 30 v
where v is the common velocity after the collision.
Since total momentum before = total momentum after,

$$60 = 30\,v$$
$$v = 2 \text{ m/s}$$

Example

A boy of mass 40 kg is travelling at 5.2 m/s when he jumps onto a stationary wagon. The mass of the wagon is 20 kg. What is the speed of the boy and the wagon after the boy jumps onto the wagon?

Solution

Before the collision, total momentum = 40 × 5.2 = 208 kg m/s
After the collision, total momentum = (40 + 20) v = 60 v
where v is the common velocity after the collision.
Since total momentum before = total momentum after,

$$60\,v = 208$$
$$v = 3.5 \text{ m/s}$$

Work and energy

Energy is a very useful quantity. Simply, energy allows objects to move and lets you do things: it is called work.

Although energy has many different forms, there are ways of measuring them. When work is done, energy is transferred to an object or changed into another form.

energy transferred = work done

= applied force × distance moved by the force

Thus $\text{Wd} = F \text{x} d$

where d distance in metres and F = applied force in newtons.

The unit of energy and work done is the joule (J); thus

$$1 \text{ joule} = 1 \text{ newton metre}$$

$$1 \text{ J} = 1 \text{ N m}$$

Example

A boy finds he has to exert a force of 50 N to lift a box 2 m onto a shelf. Calculate the work done by the boy.

Solution

$$\begin{aligned} \text{Wd} &= F \times d \\ \text{where } F &= 50 \text{ N and } d = 2 \text{ m} \\ \text{Wd} &= 50 \times 2 \\ &= 100 \text{ J} \end{aligned}$$

Example

A force of 475 N is needed to lift a crate at constant speed onto a shelf. The work done by the force is 3800 J. Calculate the height the crate is moved through.

Solution

$$\begin{aligned} \text{Wd} &= F \times d \\ \text{where } F &= 475 \text{ N and W}d = 3800 \text{ J} \\ 3800 &= 475 \times \text{d} \\ \text{d} &= 8 \text{ m} \end{aligned}$$

Example

A lorry is towed a distance of 8.5 m by a tow truck. The truck uses 6800 J to do this.
Calculate the force exerted by the tow truck on the lorry

Solution

$$\begin{aligned} \text{Wd} &= F \times d \\ \text{where } Wd &= 6800 \text{ J and } d = 8.5 \text{ m} \\ 6800 &= \text{F} \times 8.5 \\ \text{F} &= 800 \text{ N} \end{aligned}$$

Conservation of energy

Energy cannot be created or destroyed, but it can be changed from one form to another, when work is done. For example, the kinetic energy of a car is changed into mainly heat when the brakes are applied. If a system gains energy then another system loses the same amount of energy.

This is called the conservation of energy.

Gravitational potential energy (E_p)

Water in a mountain loch has stored energy (gravitational potential energy) which can be transferred into electrical energy in a hydroelectric scheme. It is available as the water is above the generating station and can be transferred by allowing it to fall. We can derive an equation to allow us to calculate the change in gravitational potential energy for a mass being lifted.

A mass m is lifted at constant speed through a vertical height of h metres (Figure 7.5). The work done in lifting it is calculated as

work done $= F \times d$

In this case the force applied must balance the weight of the box.

$$\begin{aligned}\text{applied force upwards} &= \text{weight downwards}\\ &= mg\\ \text{Work done on object} &= mg \times h\\ &= mgh\\ &= \text{gain in gravitational potential energy } (E_p)\\ E_p &= mgh\end{aligned}$$

where
E_p = change in gravitational E_p (J),
m = mass (kg),
g = gravitational field strength (N/kg) and
h = change in vertical height (m).

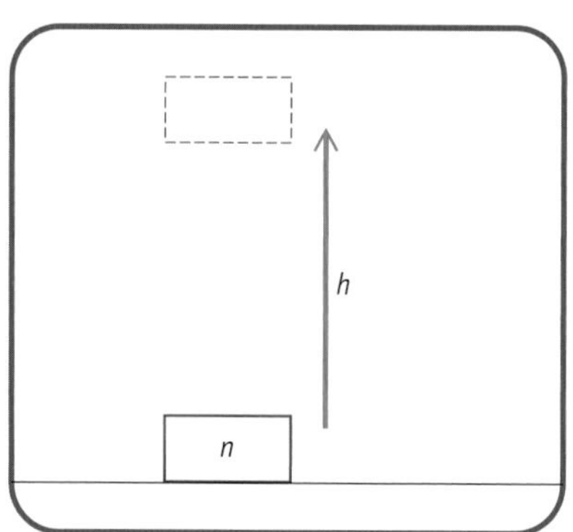

Figure 7.5 *A mass being lifted vertically at constant speed*

Kinetic Energy (E_k)

This is the energy possessed by any object which moves. We can also derive an equation for this type of energy.

A mass m starts from rest and is accelerated by a force F to a final speed v in time t (Figure 7.6).

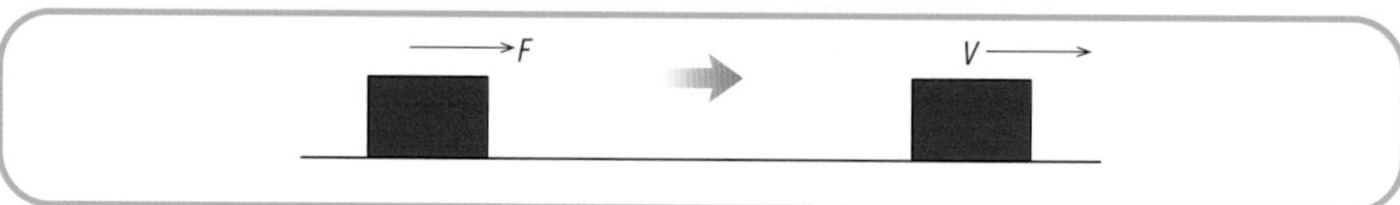

Figure 7.6 *A mass being accelerated from rest*

$$\begin{aligned}\text{Work done on object} &= F \times d\\ &= ma \times \text{area under speed–time graph (Figure 7.7)}\end{aligned}$$

$$\text{and } a = \frac{v-u}{t} = \frac{v-0}{t} = \frac{v}{t}$$

$$\text{area under the graph} = \frac{1}{2} \times t \times v = \frac{1}{2}\,vt$$

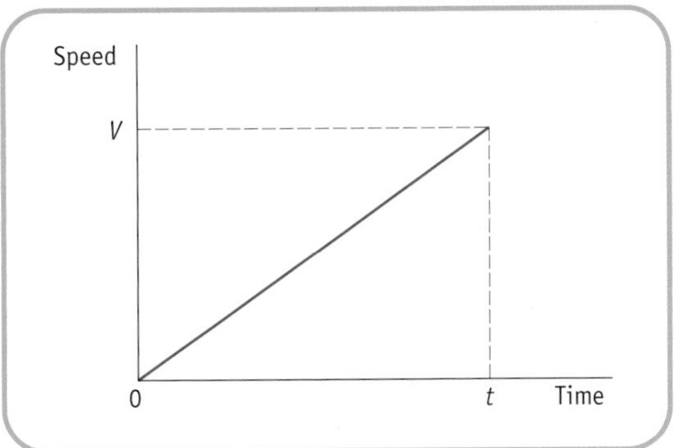

Figure 7.7 *Speed–time graph*

$$\text{work done on object} = mv \times \frac{1}{2} vt$$

$$= \frac{1}{2} mv^2$$

work done on object = gain in kinetic energy (Ek)

= big Ek − small Ek

big Ek − small Ek $= \frac{1}{2} mv^2$

but small $Ek = 0$, hence

$$Ek = \frac{1}{2} mv^2$$

Note that E_k depends on v^2. This means that:

- If the speed doubles then the kinetic energy increases by four times.
- If the speed trebles then the kinetic energy increases by nine times.

Power

When we say that one appliance is more powerful than another appliance then we mean that it uses up energy at a faster rate.

power = energy transferred in 1 s = work done in 1 s

$$\text{power} = \frac{\text{energy transferred}}{\text{time taken}} = \frac{\text{work done}}{\text{time taken}}$$

Power is measured in watts (W). A power of one watt means that one joule of energy is transferred in one second.

Example

A 3 kg box is raised through 20 m in 4 s.

a) Find the gain in gravitational E_p of the box.

b) Find the minimum power used to lift the box.

c) Why would more power be needed than the value calculated in (b)?

Solution

a) $E_p = mgh$

$= 3 \times 10 \times 20$

$= 600$ J

b) $P = \frac{E}{t}$

$= \frac{600}{4}$

$= 150$ W

c) Some of the energy supplied is changed into heat and is not used to lift the box.

Example 2

A girl of mass 50 kg is riding her bicycle. The mass of the bicycle is 15 kg. The girl and bicycle are moving at a speed of 5 m/s. The girl applies the brakes and the bicycle comes to rest in a time of 6 s.

a) What is the kinetic energy of the girl and her bicycle before she brakes?

b) What becomes of this kinetic energy during braking?

c) Calculate the minimum power of the brakes.

Solution

a) $Ek = \frac{1}{2} mv^2$

$= \frac{1}{2} \times 65 \times (5)^2$

$= 813$ J

b) It changes into heat.

c) Change in $Ek = 813$ J = energy transferred.

$P = \frac{E}{t}$

$= \frac{813}{6}$

$= 136$ W

Figure 7.8 *Estimating your own power*

Estimating your own power

In going upstairs, you will gain in gravitational potential energy.

Measure your mass on bathroom scales (Figure 7.8). To calculate your power you also need the time taken for you to go up the stairs. Run up the stairs and time how long it takes you to run up a measured vertical height.

Typical results

height of stairs = 3 m

time taken to go up stairs = 3.5 s

mass of person = 70 kg

weight = 700 N

$$
\begin{aligned}
\text{work done going up stairs} &= \text{energy transferred} \\
&= \text{gain in gravitational potential energy} \\
&= mgh \\
&= 70 \times 10 \times 3 \\
&= 2100 \text{ J}
\end{aligned}
$$

$$
\text{minimum power} = \frac{\text{energy transferred}}{\text{time taken}} = \frac{mgh}{t} = \frac{2100}{3.5}
$$

$$
= 600 \text{ W}
$$

The power of a horse, on the other hand, is about 750 W. An Olympic runner can produce up to 3000 W but we cannot sustain this power for any length of time. A typical athlete will lose up to 50% of the energy produced within the body as heat.

Energy conservation: gravitational E_p to E_k

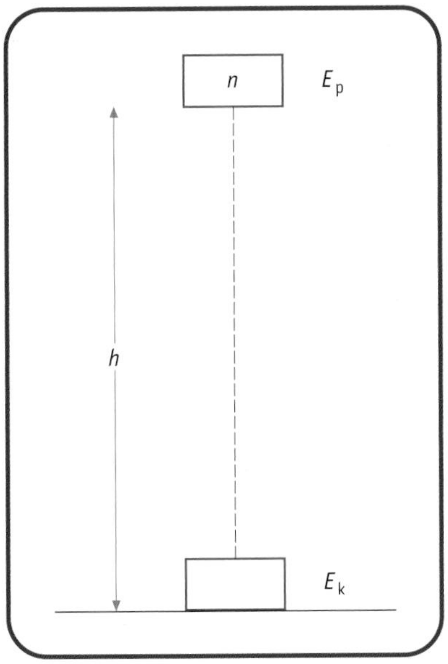

During any energy transformation the total amount of energy is always conserved; that is, it stays the same, but may be changed into less useful forms.

Energy transfer or energy changes are associated with a falling object.

As the object falls it 'loses' gravitational E_p but gains E_k. Just before impact with the ground it has 'lost' all its gravitational E_p and has only E_k (Figure 7.9).

If no energy is transferred with the surroundings, then

$$
\text{gain in } E_k = \text{loss in gravitational } Ep
$$

$$
\frac{1}{2}\text{mv}^2 - 0 = \text{mgh}
$$

$$
\frac{1}{2}\text{mv}^2 = \text{mgh}
$$

n
E_p
h
E_k

Figure 7.9 *A falling body*

Example

A tower is 750 m in height. John drops his bag over the edge. What is its speed of the bag just before reaching the ground? You may assume that the effect of of air resistance is negligible. (See Figure 7.10)

Solution

The change in gravitational potential energy equals the change in kinetic energy. Then

$$mgh = \frac{1}{2}mv^2 - 0$$

The masses will cancel on both sides of the equation, giving

$$gh = \frac{1}{2}v^2$$

The height and speed are connected but are independent of mass. All masses will fall and hit the ground at the same time if released from the same height.

$$10 \times 750 = \frac{1}{2}v^2$$

$$v^2 = 20 \times 750$$

$$= 15\,000$$

$$v = 122 \text{ m/s}$$

Figure 7.10

Efficiency

In any machine there is always some energy lost to another form of energy that is not useful to us. This occurs, for example, as work done against friction, which will produce heat in a motor car engine or any electric motor. While the total energy of the system cannot change, the useful energy output can be compared to the useful energy input. This is called efficiency.

$$\text{efficiency} = \frac{\text{useful energy output}}{\text{total energy input}} \times 100\%$$

In a similar way we can express the efficiency in terms of power:

$$\text{efficiency} = \frac{\text{useful power output}}{\text{total power input}} \times 100\%$$

Most machines have a range of efficiencies.

For a car engine the efficiency is about 20%. That is, only 20% of the energy input from the petrol is converted into kinetic energy; the remainder goes as heat either into the engine block or into the atmosphere as exhaust gases.

This may seem very low and inefficient. However we tolerate it because only petrol or diesel can give us the rapid acceleration which we demand in moving vehicles.

Example

An electric motor has an input power of 2.0 kW. The output power is 1.4 kW. Calculate the efficiency of this motor.

Solution

$$\text{Efficiency} = \frac{\text{useful power output}}{\text{total power input}} \times 100\%$$

$$= \frac{1400}{2000} \times 100$$

$$= 70\%$$

No machine can reach or exceed 100%, because this would mean it was creating energy, which is impossible, and so provides a check on your calculations.

Physics facts and key equations for momentum and energy

- Newton's third law states that forces always act in pairs which are equal in size but opposite in direction and act on separate objects.
- Momentum = mass × velocity.
- Momentum is a vector quantity and the unit is kg m/s.
- Total momentum before a collision = total momentum after, in the absence of external forces.
- Work done = force applied × distance moved by the force and is measured in joules.
- Power $P = \frac{\text{work done}}{\text{time}}$ and is measured in watts.
- Gravitational potential energy $E_p = mgh$
- Kinetic energy $Ek = \frac{1}{2} mv^2$
- $\text{Efficiency} = \frac{\text{useful work output}}{\text{total work input}} \times 100\%$

Questions

1 A boy of mass 50 kg is running at 8 m/s. He then jumps on a stationary skateboard of mass 5 kg. What is the combined speed of the boy and the skateboard?

2 A car of mass 1000 kg is travelling at 15 m/s. It collides with and sticks to a stationary lorry of mass 2500 kg.
   a) What is the momentum of the car before the collision?
   b) What is the combined velocity of the car and lorry just after the collision?

3 A full shopping trolley of mass 25 kg is being pushed at 5 m/s. It collides and locks into another trolley of mass 7 kg.
What is the combined speed of the two trolleys after collision?

4 A car and driver have a total mass of 1050 kg. The car is travelling at 21 m/s on a motorway. The car collides with a stationary car of mass 950 kg. The two cars lock together.
What is the combined speed of the two cars just after the collision?

5 In Figure 7.11 block B has a mass of 20 kg and is at rest. Block A has a mass of 10 kg and is moving at a speed of 6 m/s.
a) Calculate the momentum of block A before the collision.
b) The blocks stick together after the collision. Calculate the combined speed of the two blocks.

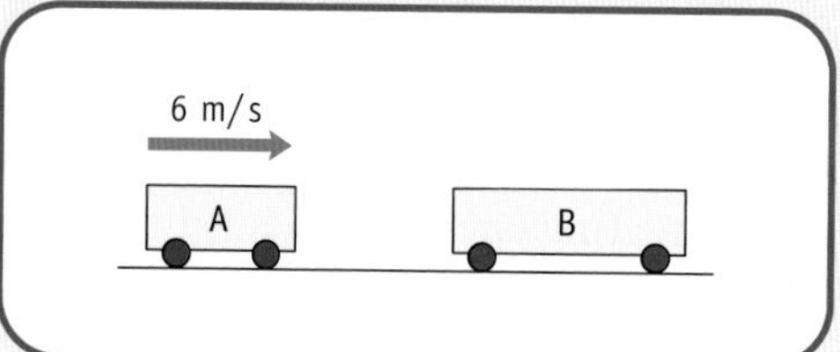

Figure 7.11

6 Copy and complete the table below.

Force (N)	Distance (m)	Work (J)
25	17	
350	8	
	12	960
	25	1200
750		15 000
20		800

7 A fork lift truck lifts a load of 5000 N at a steady speed to a height of 2.5 m.
Calculate the work done.

8 A load of 250 N is lifted at a constant speed to a height of 3 m, calculate the work done on the load.

9 Copy and complete the table below.

Mass (kg)	Gravitational field strength (N/kg)	Height raised (m)	Potential energy (J)
5	10	8	
25	10		1250
90	1.6		1440
50		2	2600

10 A wagon is moved by a force of 20 N. The wagon travels a distance of 30 m.
Calculate the work done on the wagon.

11 A force of 10 N is exerted on a supermarket trolley. The trolley moves a distance of 30m. in 5 s. Calculate the minimum power used to move the trolley.

12 Oranges hang from a branch on a tree. An orange has a mass of 200 g and is at a height of 7 m above the ground. The orange falls to the ground.
Calculate the change in gravitational potential energy as it reaches the ground.

13 A student of mass 50 kg climbs a set of stairs. Each step is 0.2 m high and there are 18 steps.
a) Calculate the gravitational potential energy gained by the student
b) The student climbs the stairs in 4 s. Calculate the minimum power transferred by the student.

14 A model rocket is fired straight up at an initial speed of 8 m/s. The rocket has a mass of 0.2 kg.
a) Calculate the intitial kinetic energy of the rocket.
b) The mass of the rocket does not change. The rocket reaches its maximum height. What is the gravitational potential energy gained by the rocket?
c) Use your answer from b) to calculate the height that the rocket reached.

15 A ball of mass 0.5 kg is dropped from a tower which is 75 m high.
a) When the ball reaches the ground, calculate the change in gravitational potential energy of the ball.
b) Assuming all of this energy is transferred into kinetic energy, calculate the speed of the ball just before it reaches the ground.

16 A car is being driven along a road at 15 m/s. The total mass of the car and driver is 900 kg.
a) Calculate the kinetic energy of the car and driver.
b) The brakes are applied and the car is brought to rest. What is the work done by the brakes?
c) If the distance travelled in braking is 200 m, calculate the average force exerted by the brakes.

17 A crane lifts a steel bar of mass 500 kg to a height of 8 m above the ground.
a) Calculate the gravitational potential energy gained by the bar.
b) The lifting operation is done in 4s. Calculate the minimum power of the crane engine.
c) Why will the actual power of the crane engine be greater than the value calculated in b)?

18 A rocket of mass 20 000 kg is travelling vertically upwards at a speed of 1000 m/s. Calculate the kinetic energy of the rocket.

19 Copy and complete the table below.

Mass (kg)	Speed (m/s)	Kinetic Energy (J)
5	10	
20		2250
	13	1014

20 A ball of mass 0.25 kg falls from a height of 3.2 m.
   a) Calculate the change in gravitational potential energy when the ball reaches the ground.
   b) Assuming that all the gravitational potential energy is transferred to kinetic energy just before the ball reaches the ground, calculate the speed of the ball on reaching the ground.

21 A motor lifts a mass of 8 kg through a height of 6 m. This takes a time of 4 s to complete. The power rating of the motor is 200 w.
   a) Calculate the efficiency of the lifting operation.
   b) Suggest why the efficiency is not 100%.

22 A ball of mass 3 kg falls through a height of 4 m onto the ground.
   a) What is the change in gravitational potential energy of the ball?
   b) Just before reaching the ground the speed of the ball is 8 m/s. Calculate the kinetic energy of the ball.
   c) Calculate the efficiency of the energy transfer to the ball.

Heat

At the end of this chapter you should be able to...

1 State that the same mass of different materials need different quantities of heat energy to change their temperature by one degree celsius.
2 Carry out calculations involving specific heat capacity.
3 State that heat is gained or lost by a substance when its state is changed.
4 State that a change of state does not involve a change in temperature.
5 Carry out calculations involving specific latent heat.
6 Carry out calculations involving energy, work, power and the principle of conservation of energy.

Heat and temperature

Heat, just like light and sound, is a form of energy and is measured in joules (J). Temperature is a measure of how hot a substance is and is measured in degrees celsius (°C).

Heat transfer

Heat is always transferred from a higher temperature to a lower temperature. There are three possible ways in which heat can be transferred called conduction, convection and radiation.

Conduction

The heat is transferred through a solid material. The particles making up the solid cannot change their position but pass the heat from particle to particle. Heat moves from the high temperature to the low temperature. Materials which allow heat to move easily through them are called conductors – metals are the best conductors. Materials which do not allow heat to move through them easily are called insulators – liquids and gases are good insulators (poor conductors).

Convection

The heat is transferred by the movement of the heated particles making up the liquid or gas. The heated fluid (liquid or gas) becomes less dense and rises, cold fluid falls to take its place, and convection currents are set up. Convection cannot take place in a solid because the particles cannot move away.

Many heat insulators contain trapped air, e.g. cotton wool, felt, woollen clothes. No convection can take place because the air is trapped and cannot move. Air is also a non-metal, so it is a poor conductor of heat.

Radiation

Radiation travels in straight lines until absorbed by an object. It can travel through a vacuum (the Earth is heated by radiation from the sun). Heat radiation travels at a speed of 3×10^8 m/s. All hot materials radiate heat energy.

> **Did you know?**
>
> All houses lose heat by conduction, convection and radiation to the colder outside. To have a comfortable living temperature in the house but avoid high fuel bills it is important to reduce these heat losses. The main heat losses from a house, and methods of prevention, are shown in Figure 8.1.

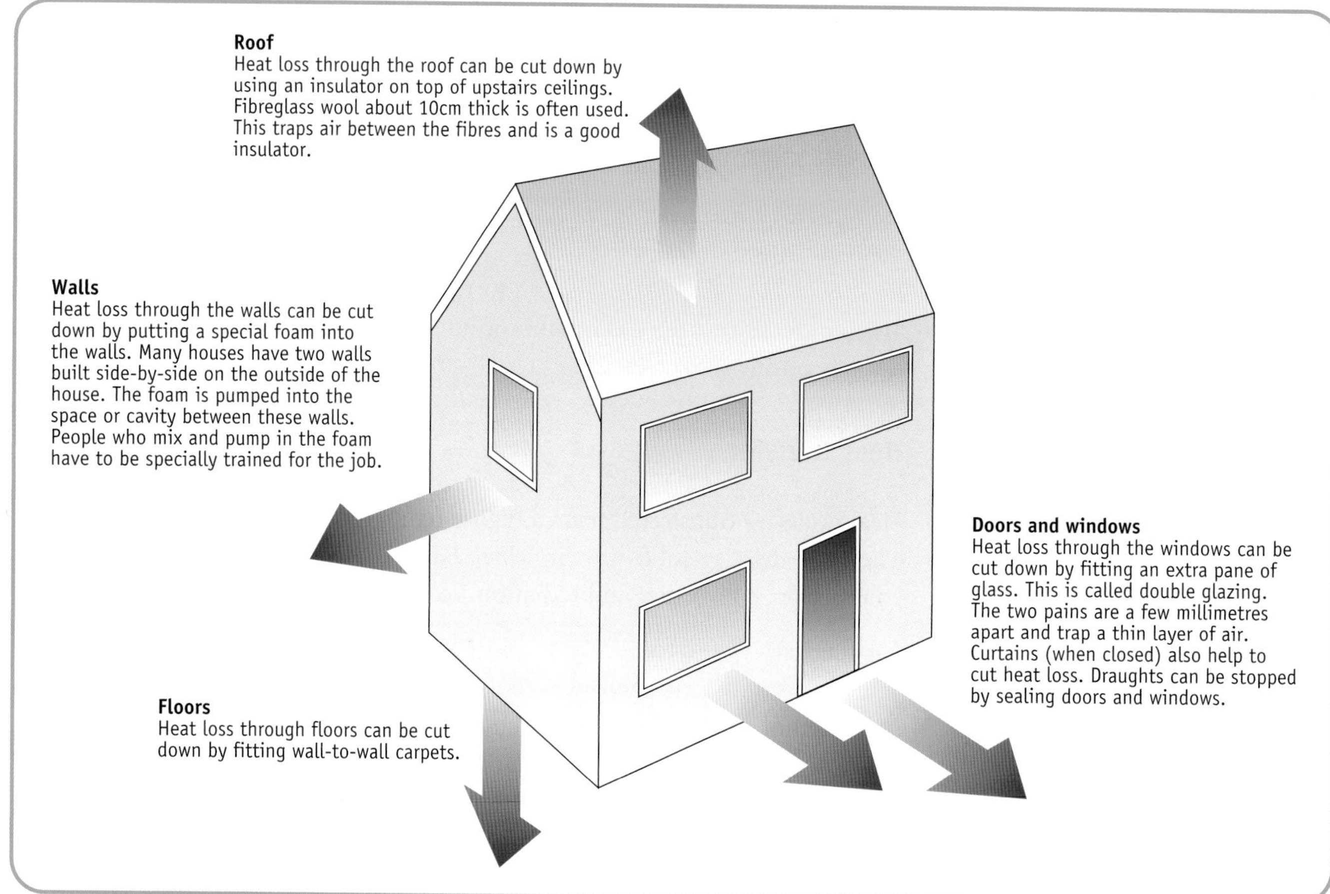

Figure 8.1 *Main heat losses from a house*

> **Did you know?**
>
> Warm objects give off invisible 'heat rays' called infrared radiation. These rays are invisible to our eyes but can be viewed using an infrared camera as shown in Figure 8.2. In some cameras, colour photographs showing different temperatures are produced. These pictures are called thermograms and can be particularly useful in medicine. For instance, Figure 8.3 shows the thermogram

of an arthritic elbow of a patient. The orange/yellow colours on this thermogram show abnormally high temperatures indicating inflammation of the elbow joint.

Figure 8.2 *Firefighters use infrared cameras*

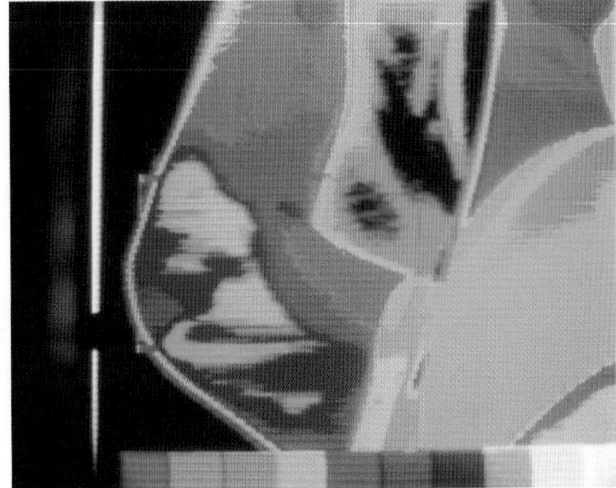

Figure 8.3 *Thermogram*

Physiotherapists also use infrared radiation to penetrate the skin and to heat damaged muscles. Heat causes the muscles to heal more rapidly.

Changing temperature

When you want to make a hot drink you put some water in the kettle and switch the kettle on. The water will get hot as a result of the energy supplied by the heating element of the kettle. However, the amount of energy required to warm the water depends on:

- The temperature rise – more energy is needed for a larger temperature rise.
- The mass of the water – more energy is needed for a greater mass of water.

Example

It requires 20 900 J of energy to increase the temperature of 0.5 kg of water by 10°C.

a) How much energy is required to increase the temperature of 0.5 kg of water by 20°C?

b) How much energy is required to increase the temperature of 1 kg of water by 40°C?

Solution

a) It takes 20 900 J to change the temperature of 0.5 kg of water by 10°C, so it will take twice as much energy to change the temperature of 0.5 kg by 20°C. Therefore it takes 41 800 J to change the temperature of 0.5 kg of water by 20°C.

b) It takes 20 900 J to change the temperature of 0.5 kg of water by 10°C.
It takes 41 800 J to change the temperature of 1 kg of water by 10°C.
It takes 167 200 J to change the temperature of 1 kg of water by 40°C.

Specific heat capacity

When equal masses of water and copper are supplied with the same quantity of heat, it is found that the copper has a higher rise in temperature.

The energy needed to change the temperature of a material depends on:

- The change in temperature of the material (ΔT).
- The mass of the material (m).
- The specific heat capacity of the material, c.

The specific heat capacity of a material is the amount of energy required to change the temperature of 1 kg of the material by 1°C. The unit of specific heat capacity is joules per kilogram per degree celsius (J/kg °C).

Water has a specific heat capacity of 4180 joules per kilogram per degree celsius (4180 J/kg °C). This means that it takes 4180 joules of energy to change the temperature of 1 kg of water by 1°C. It would take 16 720 J to change the temperature of 4 kg of water by 1°C, and it would take 83 600 J to change the temperature of 4 kg of water by 5°C.

The energy needed to change the temperature of a material can be put in the form of an equation:

$$E_h = cm\Delta T$$

where E_h = energy needed to change the temperature of the material (J);

c = specific heat capacity of the material (J/kg °C);

m = mass of the material (kg);

and ΔT = change in temperature of the material (°C).

Example

A 750 g steel cooking pan cools from 65°C to 18°C. How much heat energy is released by the pan? The specific heat capacity of steel is 500 J/kg °C.

Solution

$E_h = cm\Delta T$

$E_h = 500 \times 0.750 \times (65 - 18)$

$E_h = 500 \times 0.750 \times 47$

$E_h = 1.76 \times 10^4$ J

Example

An insulated container contains 0.5 kg of water at an initial temperature of 16°C. A heater is immersed in the water and switched on. The heater supplies 83 600 J of energy to the water.
Calculate the final temperature of the water. The specific heat capacity of water is 4180 J/kg °C.

Solution

$E_h = cm\Delta T$

$83\,600 = 4180 \times 0.5 \times \Delta T$

$83\,600 = 2090 \times \Delta T$

$$\Delta T = \frac{83\,600}{2090} = 40°\text{C}$$

But ΔT = Final temperature − Initial temperature

40 = Final temperature − 16

Final temperature = 40 + 16 = 56°C

Heat problems

Energy can be changed from one form to another. The total amount of energy remains unchanged – this is the principle of conservation of energy. An electric heater converts electrical energy into an equal amount of heat.

Due to conduction, convection and radiation, some of the heat supplied by the heater will be transferred ('lost') to the surroundings. This means that the material will absorb (take in) less energy than was supplied by the heater.

energy supplied = energy absorbed + energy transferred to the surroundings

In most heat problems it is assumed that no energy is transferred to the surroundings. Hence:

energy supplied = energy absorbed by the material

Example

A well-insulated kettle contains 0.8 kg of water. The kettle is switched on for 2 minutes. The temperature of the water changes from 16°C to 100°C. Calculate the power rating of the element of the kettle. The specific heat capacity of water is 4180 J/kg °C.

Solution

$E_h = cm\Delta T = 4180 \times 0.8 \times (100 - 16)$

$E_h = 4180 \times 0.8 \times 84 = 280\ 896\ \text{J}$

$$P = \frac{E}{t} = \frac{280\ 896}{2 \times 60} = 2341\ \text{W} = 2.3\ \text{kW}$$

Example

A deep fat fryer is used to heat 800 g of cooking oil. The temperature of the cooking oil changes from 20°C to 140°C in a time of 180 s. The specific heat capacity of the cooking oil is 3000 J/kg °C.

a) Calculate the power rating of the element of the deep fat fryer.
b) The deep fat fryer operates from the 230 V mains. Find the current in the element of the deep fat fryer.

Solution

a) $E_h = cm\Delta T = 3000 \times 0.8 \times (140 - 20)$

$E_h = 3000 \times 0.8 \times 120 = 288\ 000\ \text{J}$

$$P = \frac{E}{t} = \frac{280\ 000}{180} = 1600\ \text{W} = 1.6\ \text{kW}$$

b) $P = IV$

$1600 = I \times 230$

$$I = \frac{1600}{230} = 6.96\ \text{A}$$

Specific latent heat

When cold water in a kettle is heated its temperature rises until the water starts to boil at 100°C. Further heating of the water no longer produces a rise in temperature of the water but steam at 100°C is produced. The energy supplied by the kettle is now being used to change water at 100°C into steam at 100°C. The energy required to change 1 kg of a liquid at its boiling point into 1 kg of vapour at the same temperature is called the specific latent heat of vaporisation. The word **latent** means hidden and refers to the fact that the temperature of the material does not change and the energy supplied to the material seems to have disappeared.

When ice at its melting point of 0°C is heated it turns into water at 0°C. Energy is required to change the ice to water without a change in temperature. The energy required to change 1 kg of a solid at its melting point into 1 kg of liquid at the same temperature is called the specific latent heat of fusion.

The three states of matter are solid, liquid and gas. Whenever a material changes state, latent heat is required.

When a material changes from a solid to a liquid or a liquid to a gas, energy is needed to break down the force (or bond) holding the particles together and to push the particles further apart.

- Specific latent heat of fusion: the energy required to change 1 kg of a solid at its melting point to a liquid without change in temperature.
- Specific latent heat of vaporisation: the energy required to change 1 kg of a liquid at its boiling point to a gas without change in temperature.

When a material changes state from solid to liquid or liquid to gas, latent heat is absorbed (taken in). When a material changes state from gas to liquid or liquid to solid, latent heat is released (given out). When a material changes state there is no change in temperature.

The symbol for specific latent heat is l. Specific latent heat is measured in J/kg.

The specific latent heat of a material is the energy required to change the state of 1 kg of the material without a change in temperature.

For a material with a specific latent heat l:

- To change the state of 1 kg of the material at constant temperature requires l J.
- To change the state of m kg of the material at constant temperature requires $m \times l$ J.

That is, $E_h = ml$

where E_h = energy needed to change state of material (J);

m = mass of material which changed state (kg);

l = specific latent heat of material (J/kg)

Example

Ice cubes at 0°C are placed in a well-insulated container. The mass of the ice is 0.6 kg. A heater with a power rating of 50 W is now placed in the ice in the container. How long does the heater require to be switched on to melt 0.05 kg of the ice? (Assume that no ice is melted due to heat from the room and that all the energy suppied by the heater is absorbed by the ice.) The specific latent heat of fusion of water is 3.34×10^5 J/kg.

Solution

energy required to melt ice $= ml$

$= 0.05 \times 3.34 \times 10^5$

$= 1.67 \times 10^4$ J

energy supplied by heater = energy required to melt ice

$P \times t = 1.67 \times 10^4$

$50 \times t = 1.67 \times 10^4$

$$t = \frac{1.67 \times 10^4}{50} = 334 \text{ s}$$

Example

A student uses a heater to boil water in a container. When the water is boiling the following information is recorded:
mass of water changed to steam = 100 g;
power rating of heater = 2200 W;
time taken to change 100 g of water into steam = 105 s.

a) Calculate the specific latent heat of vaporisation of water obtained from this experiment.
b) Explain why this value is likely to be higher than the accepted value for the specific latent heat of vaporisation of water.

Solution

a) Energy supplied by heater $E_h = P \times t = 2200 \times 105 = 231\ 000$ J.

$$\text{Also} \quad E_h = ml$$
$$231\ 000 = 0.1\ l$$
$$l = \frac{0.1}{231\ 000} = 2\ 310\ 000 \text{ J/kg}$$

b) Not all of the energy supplied by the heater is absorbed by the water. Some of it is 'lost' in heating up the air around the heater. This means that more energy is supplied by the heater to produce 0.1 kg of steam than is actually required. The makes the value for l too large.

Did you know?

Latent heat is used to help keep food in a Cool Box, the type used for picnics, cool.
Special chemical packs are placed in a freezer. The chemicals are a liquid at normal room temperature but change to a solid during cooling in a freezer. The frozen chemical packs are placed on top of the food in the Cool Box – air surrounding the frozen packs cools and falls due to convection. Initially the cold frozen packs keep the box cool, as they take heat from the food. However, the temperature in the Cool Box will slowly rise and the chemicals will begin to melt i.e. change from a solid to a liquid. The energy needed to bring about this change in state is absorbed from the food. The food 'loses' heat and so is kept cool for a longer period.

Did you know?

At normal atmospheric pressure, water boils at 100°C. However, at pressures higher than atmospheric pressure, water boils at a higher temperature. The steam produced will be at a temperature greater than 100°C. In a pressure cooker, due to the higher pressure, the temperature of the steam is about 115°C. The steam is forced through the food, which cooks more rapidly as a result of the higher temperature. This saves energy.

Physics facts and key equations for heat

- Equal masses of different materials require different amounts of energy to change their temperature by 1°C.
- Energy needed to change temperature = specific heat capacity × mass × change in temperature:

$$E_h = cm\Delta T$$

  This equation is used whenever there is a change in temperature of the material.
- During a change in temperature the energy absorbed or lost by a material, E_h, is measured in joules (J), specific heat capacity, c, is measured in joules per kilogram per degree celsius (J/kg °C), mass, m, is measured in kilograms (kg) and the change in temperature, ΔT, is measured in degrees celsius (°C).
- A specific heat capacity, c, of 100 J/kg °C means that 100 J of energy are required to change the temperature of 1 kg of the material by 1°C.
- A change of state occurs when a solid changes into a liquid (or a liquid changes to a solid) or a liquid changes to a gas (or a gas changes to a liquid).
- There is no change in temperature when a change of state occurs.
- Energy needed to change state = mass x specific latent heat

$$E_h = ml$$

  This equation is used whenever there is a change in state of the material.
- During a change in state, the energy absorbed or lost by the material, E_h, is measured in joules (J), mass, m, is measured in kilograms (kg) and the specific latent heat of fusion or vaporisation, l, is measured in joules per kilogram (J/kg).

Questions

1 It takes 8360 J of energy to raise the temperature of 2 kg of water by 1°C. How much energy will be required to raise the temperature of:
a) 4 kg of the water by 1°C; b) 4 kg of the water by 5°C; c) 8 kg of the water by 10°C?

2 Calculate the amount of energy required to raise the temperature of
a) 0.5 kg of water from 18°C to 58°C (the specific heat capacity of water is 4180 J/kg °C).
b) a 0.95 kg steel baking tray from 18°C to 198°C (the specific heat capacity of steel is 500 J/kg °C).

3 A well-insulated kettle contains 1.2 kg of water at a temperature of 20°C. The kettle is switched on. It takes 180 s for the water to reach a temperature of 100°C.
The specific heat capacity of water is 4180 J/kg °C
a) How much energy is absorbed by the water in 180 s?
b) Calculate the power rating of the heating element of the kettle.

4 A heater is connected to a 12 V supply. When the heater is operating, the current in the heater is 4 A. The heater is used to heat a 1 kg copper block. The heater is switched on for 5 minutes (the specific heat capacity of copper is 386 J/kg °C).

a) How much heat was produced by the heater in 5 minutes?

b) Calculate the maximum possible rise in the temperature of the copper block.

c) Explain why the rise in temperature of the copper block will be less than the answer to (b).

5 A kettle contains 0.4 kg of water at an initial temperature of 18°C. The element of the kettle is rated at 2000 W. The kettle is switched on for 40 s. The specific heat capacity of water is 4180 J/kg °C.

a) How much energy was produced by the element of the heater in 40 s?

b) Calculate the maximum final temperature of the water.

6 How much energy is required to change 0.8 kg of ice at 0°C into water at 0°C? (the specific latent heat of fusion of ice is 3.34×10^5 J/kg).

7 How much energy is required to change 0.2 kg of steam at 100°C into water at 100°C? (the specific latent heat of vaporisation of water is 2.26×10^6 J/kg).

8 A heater rated at 2000 W heater is used to bring 1.5 kg of water to its boiling point. The heater is left on for a further 80 s after the water has reached its boiling point. Calculate the mass of water changed to steam in this time. The specific latent heat of vaporisation of water is 2.26×10^6 J/kg.

Exam Questions

1 a) A car accelerates from rest along a straight road for 20s. The car reaches a speed of 15 m/s. The car continues at this speed for 30 s and then comes to a halt in a further 10 s.
   (i) Sketch a speed–time graph for the entire journey.
   (ii) Calculate the acceleration of the car.
   (iii) Calculate the deceleration of the car.
   (iv) Calculate the distance travelled by the car in the last 10 s of its journey.

b) A shopping trolley of mass 12 kg is pushed along a floor. The force exerted by the shopper on the trolley is 32 N. The frictional force acting on the trolley is 14 N.
   (i) Show that the acceleration of the trolley is 1.5 m/s^2.
   (ii) The trolley starts from rest. Calculate the speed of the trolley after 8 s.
   (iii) The trolley is pushed on to a rough surface of a car park. State and explain what will happen to the acceleration of the trolley.

2 A ball is kicked from a cliff with a horizontal speed of 15 m/s. It reaches the ground 3.5 s later.
a) Calculate the horizontal distance travelled.
b) Calculate the vertical speed of the ball just before it hits the ground.
c) Another identical ball is kicked with a horizontal speed of 18 m/s. How does
   (i) the horizontal distance compare with the value calculated in a)?
   (ii) The vertical speed just before hitting the ground compare with the value calculated in b)?

3 A car of mass 1000 kg is travelling along a straight road at a constant speed of 15 m/s. It collides with a stationary car of mass 1200 kg.
a) During the collision momentum is conserved. Explain what is meant by this statement.
b) The two cars lock together after the collision. Calculate their combined speed.
c) Velocity is a vector. Explain what is meant by this term.
d) A passenger in the first car is wearing a seat belt. Explain in terms of Newton's laws of motion how this protects the passenger.

4 An ice tray contains 0.15 kg of water. The initial temperature of the water is 20°C. The ice tray and the water are placed in a freezer. The freezer is at a constant temperature of −19°C.
Show that the freezer has to remove 68.6 kJ of heat energy in order to change 0.15 kg of water at 20°C into ice at −19°C (the specific heat capacity of water = 4180 J/kg °C, the specific latent heat of fusion of ice = 3.34×10^5 J/kg, the specific heat capacity of ice = 2100 J/kg °C).

5 The element of an immersion heater is completely immersed in 1.2 kg of water. The heater is connected to a 12 V supply and switched on. The temperature of the water rises by 12.5°C in 23 minutes. The specific heat capacity of water is 4180 J/kg °C.
a) Calculate the amount of energy gained by the water.
b) Calculate the power rating of the heater, stating any assumption you make. (Use an appropriate number of figures in your answer.)
c) Find the current in the element of the heater.

6 A sample of a solid is at a temperature of 20°C. The mass of the sample is 0.4 kg. A heater, rated at 100 W, is used to heat the sample for 1100 s. Figure E.1 shows how the temperature of the sample varies with time.
a) What is the melting point of the sample?
b) Calculate the specific heat capacity of the sample in the solid state.
c) Calculate the specific latent heat of fusion of the sample.

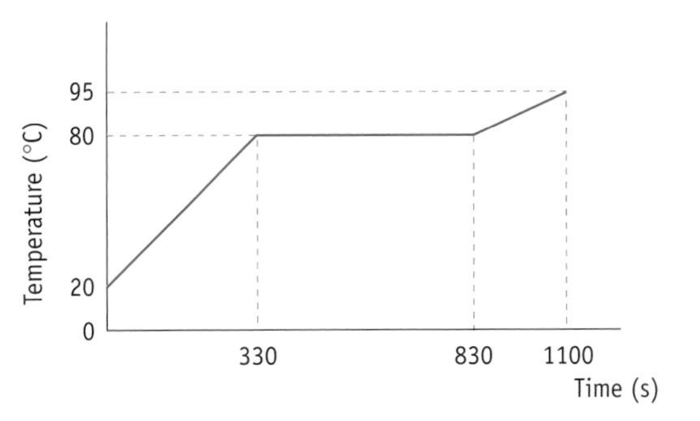

Figure E.1

Waves and Optics

Waves and optics are an important part of communication technology. As well as radio, television and mobile phones, they are used to transmit information across the world in a fast and reliable way. As well as communicating, waves and optics are used widely in medical examinations, operations and treatments.

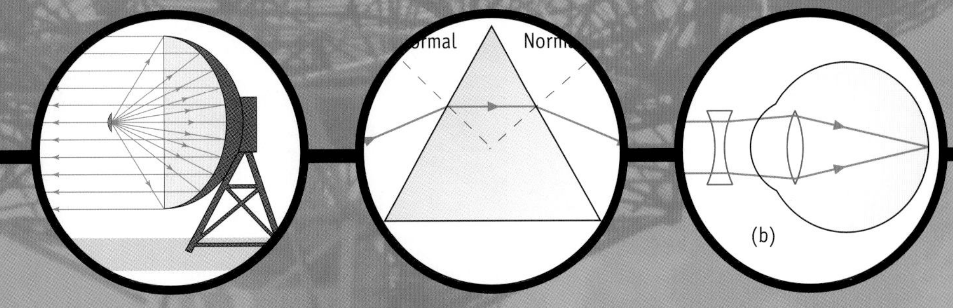

9 Waves

At the end of this chapter you should be able to...

1 State that a wave transfers energy.
2 Describe a method of measuring the speed of sound in air, using the relationship between distance, time and speed.
3 State that radio and television signals are transmitted through air at 300 000 000 m/s and that light is also transmitted at this speed.
4 Carry out calculations involving the relationship between distance, time and speed in problems on water waves, sound waves, radio waves and light waves.
5 Use the following terms correctly in context: wave, frequency, wavelength, speed, amplitude and period.
6 State the difference between a transverse and a longitudinal wave and give an example of each.
7 Carry out calculations involving the relationship between speed, wavelength and frequency for waves.
8 State, in order of wavelength, the members of the electromagnetic spectrum: gamma rays, X-rays, ultraviolet, visible light, infrared, microwaves, TV and radio.

Waves

Waves such as water waves are easily visible. We can see them breaking on the shore and the effect of large waves can cause considerable damage during storms (Figure 9.1). In 2004 a tsunami damaged parts of the Far East and in

Figure 9.1 *Damage caused by hurricane Katrina in New Orleans*

2005 the damage by hurricane Katrina devastated New Orleans due to the dams bursting. These natural events show that the energy caused by large waves is considerable.

Waves transfer energy from one point to another. In the case of water waves this transfer can be seen since wood and other materials are often washed onto beaches. However, sound, light, radio and TV waves are also methods of transferring energy.

We see a lightning flash before we hear the thunder. This tells us that light travels faster than sound, but how fast does sound travel? To answer this question we need to measure the speed of sound.

Measuring the speed of sound

This requires the equation met earlier that average speed $= \frac{\text{distance}}{\text{time}}$

$$v = \frac{d}{t}$$

In a material the speed of sound is constant. So the average speed is the speed of sound in the material.

One method is to use a source of sound such as a pair of cymbals and a stopwatch and measuring tape.

A student stands at one of a field with a pair of cymbals.

Another student stands at the far end of the field with a stopwatch.

The distance (d) between the two students is measured in metres with the measuring tape.

The stopwatch is started when the timekeeper sees the cymbals coming together.

The stopwatch is stopped when the timekeeper hears the sound of the cymbals and the time (t) is noted (Figure 9.2).

The equation $v = \frac{d}{t}$ is now used.

Figure 9.2 *Measuring the speed of sound using cymbals and stopwatch.*

An accurate method uses a computer as a timer to measure the time for a sound signal to travel from one microphone to another (Figure 9.3). This avoids reaction time which is important if the time interval is very small.

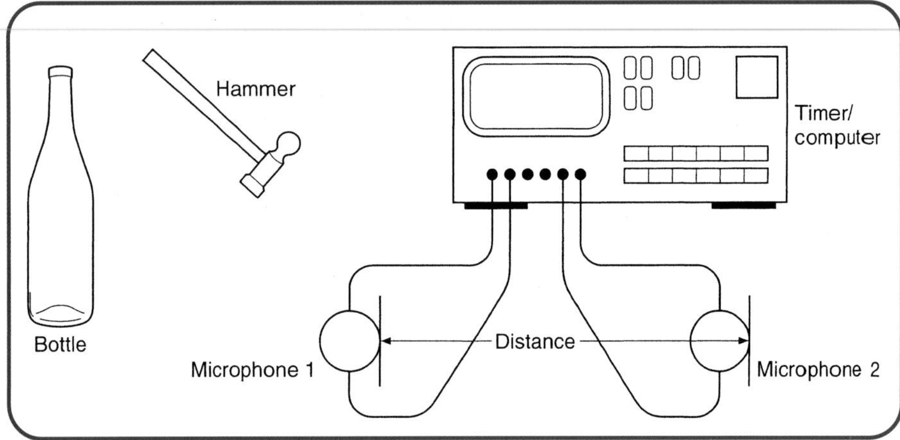

Figure 9.3A *Measuring the speed of sound*

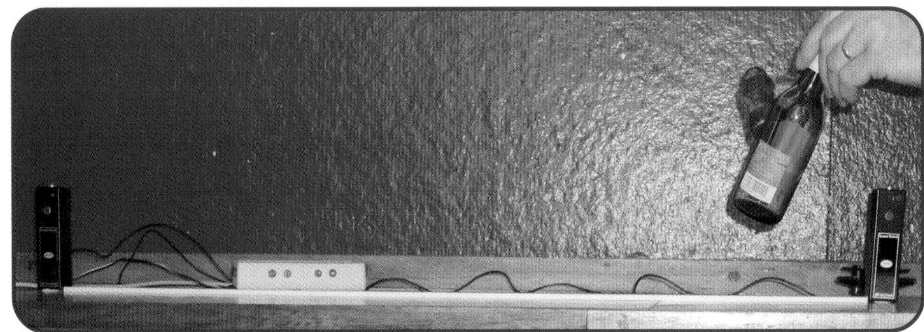

Figure 9.3B

The microphones are placed a known distance apart. The distance will be measured precisely with a metre rule.

- When a loud sound is made by hitting a bottle, the first microphone starts the timing.
- When the sound reaches the second microphone the timer stops timing.

Typical results

distance between the microphones = 2.0 m

time on the timer = 6 ms = = 0.006 s (ms is a millisecond, which is 1/1000 s)

$$\text{speed of sound} = \frac{\text{distance travelled}}{\text{time taken}} = \frac{2.0}{0.006}$$

value of speed of sound = 333 m/s

In air the speed of sound will vary due to temperature and the value given in the data table is 340 m/s.

The speed of light

Radio and television signals travel about one million times faster than sound. These waves and light waves travel in air at 300 million m/s. This value for the speed of light is now the constant against which all other physical measurements can be found.

Properties of waves

All waves have certain properties, which are shown in Figure 9.4. The top part of the wave is called the **crest** and the bottom part is the **trough**. The line running through the middle of the wave pattern is called the **axis**.

The distance from the axis to the top of the crest or the bottom of the trough is called the **amplitude**. The **wavelength** is the distance after which the pattern repeats itself. It is given the symbol λ (lambda).

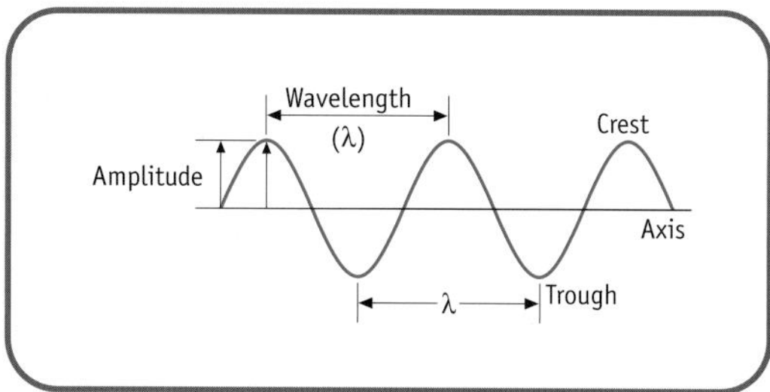

Figure 9.4 *Properties of a wave*

All distances are measured in metres or occasionally millimetres.

The **frequency** of the wave is the number of waves per second and is measured in hertz (Hz).

$$\text{Frequency} = \frac{\text{number of waves produced}}{\text{time taken}}$$

The **period** is the time for one complete wave to travel past a point and is measured in seconds.

The link between frequency f and period T is given as

$$T = \frac{1}{f}$$

If the frequency is known the period can be found and vice versa.

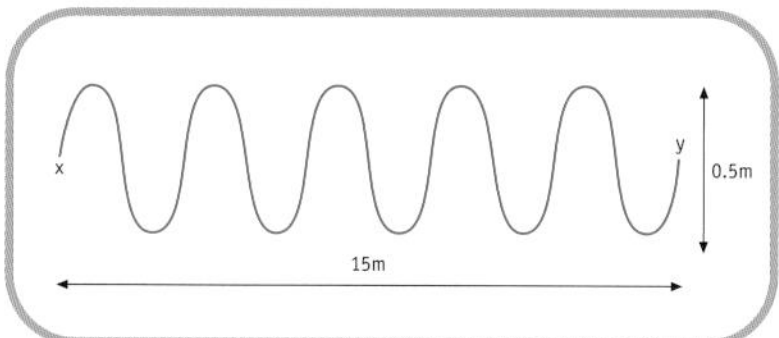

Figure 9.5

Example

The waves shown in Figure 9.5 takes 5 s to go from X to Y. The distance from X to Y is 15 m.

a) Calculate the speed of the wave.

b) What is the wavelength of this wave?

c) What is the amplitude of the wave.

Solution

a) Using the equation $v = \frac{d}{t}$

$$v = \frac{15}{5}$$

$$= 3 \text{ m/s}$$

b) 5 complete waves occupy the distance XY which is 15 m

one wave occupies a distance of $\frac{15}{5} = 3$ m

One wavelength is 3 m

c) Amplitude is the height from axis to top of crest which is 0.25 m

Example

64 waves reach a shore in a time of 8 s.

Calculate a) the frequency of the wave
b) the period of the wave

Solution

a) $\text{Frequency} = \frac{\text{number of waves produced}}{\text{time taken}}$

$= \frac{64}{8}$

$= 8$ Hz

b) $T = \frac{1}{f}$

$= \frac{1}{8}$

$= 0.125$ s

Example

A light signal is sent a distance of 5 km in air. Calculate the time for this journey.

Solution

Using the equation $v = \frac{d}{t}$

$v = 300\,000\,000$ m/s and $d = 5000$ m

$t = \frac{5000}{300\,000\,000}$

$= 1.7 \times 10^{-5}$ s

This very small time shows that if TV and radio are broadcast at this speed then we can receive instantaneous pictures of events from around the world.

Transverse and longitudinal waves

If we take a slinky spring then it is possible to create two different types of waves.

If you push the end of the spring back and forth then the coils become alternately compressed and then extended. You can also see the wave running along the coil. This type of wave is called a longitudinal wave (Figure 9.6). In this type of wave the particles of the material (medium) vibrate back and forwards along the direction the wave is travelling.

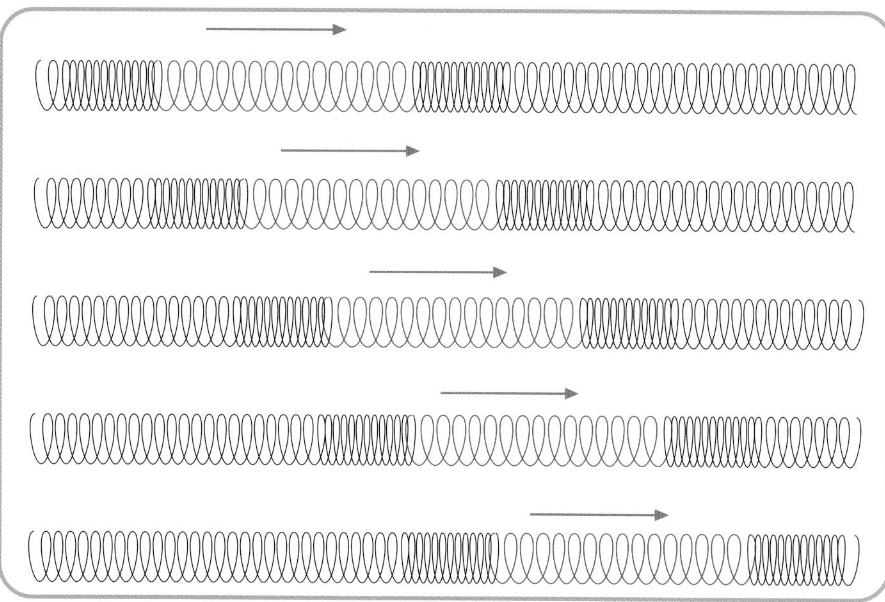

Figure 9.6 *The **pattern of disturbance** is in the **same direction** as the **direction of wave movement**.*

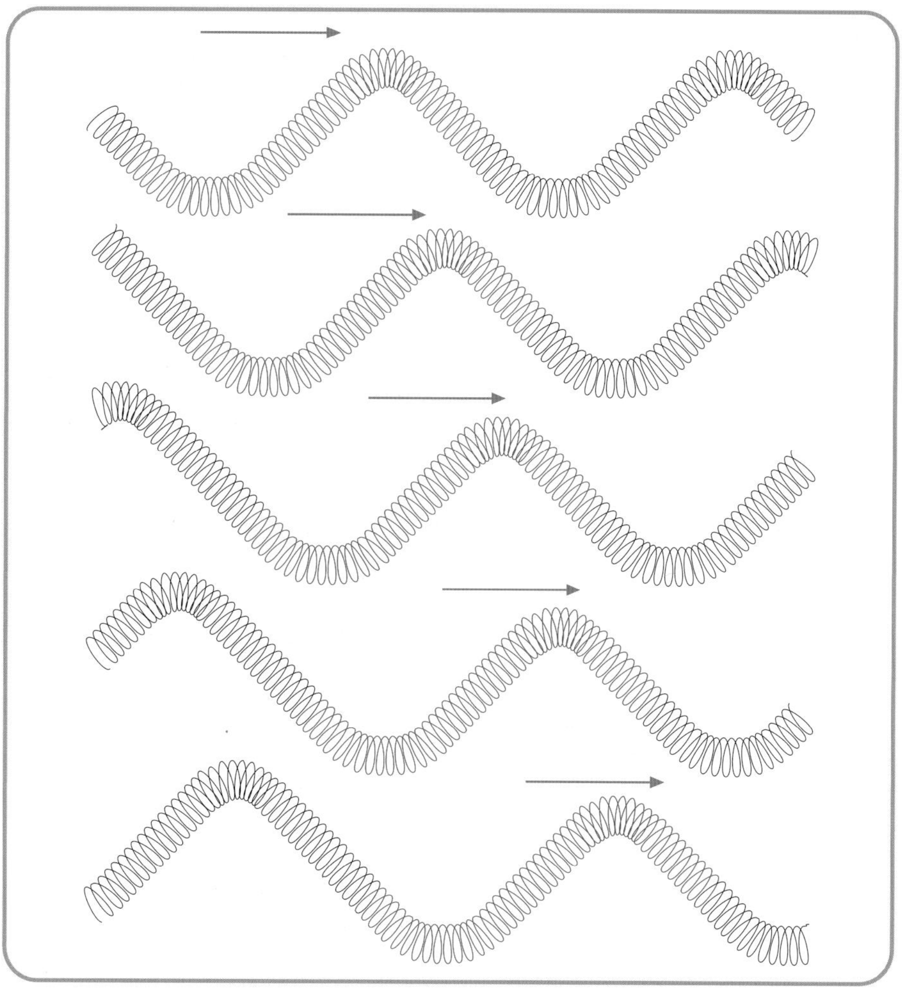

Figure 9.7 *The **pattern of disturbance** is at **right angles** (90°) to the **direction of wave movement**.*

If you move the end of the spring from side to side then each coil moves from side to side as the wave travels along the spring (Figure 9.7). This is called a transverse wave, in which the particles of the medium vibrate at right angles to the direction of motion of the wave.

- Sound waves are longitudinal waves.
- Light and all forms of electromagnetic waves are transverse waves.

Did you know?

Some waves can be made of the two different types of waves. Seismic waves produced from the Earth's core and water waves are complex forms of both transverse and longitudinal waves.

Tsunami is a seismic sea wave caused by an earthquake occuring less than 30 km below the seafloor with a magnitude of a major earthquake. It can have a wavelength of more than 100 km and a time period of one hour. In the Pacific Ocean the water depth is about 4 km and the wavelength of a tsunami is about 50 km. The speed of such a wave is about 200 m/s or 700 km/hr. In the deep ocean this type of wave has an amplitude of only about a metre but as it approaches the coastline, its speed decreases and its amplitude increases. This large increase in amplitude causes the large damage when it hits buildings (Figure 9.8).

Figure 9.8 *A tidal wave hits Penang following the Asian tsunami (December 26 2004)*

Speed of waves

The speed of the wave can be found from the usual equation:

$$\text{average speed} = \frac{\text{distance travelled}}{\text{time taken}}$$

$$v = \frac{d}{t}$$

where v = speed in metres per second (since in a given material the speed of a wave does not change), d = distance travelled in metres and t = time taken in seconds.

The speed of the wave can also be found from the following equation. In words:

speed = frequency ∞ wavelength

$$v = f \times \lambda$$

In units: m/s = hertz x metres

This means that there are two equations for finding the speed of waves.

Wave calculations

We can use the wave equations to find some of the features of waves.

Example

Waves of frequency 6 Hz are sent down a rope. The speed of the waves is 12 m/s.
Calculate the wavelength and the period of the wave.

Solution

Using $v = f \times \lambda$

$f = 6$ Hz and $v = 12$ m/s

$$\lambda = \frac{v}{f} = \frac{12}{6} = 2 \text{ m}$$

$$T = \frac{1}{f} = \frac{1}{6} = 0.17 \text{ s}$$

Example

A radio station broadcasts on a frequency of 198 kHz. What is the wavelength of this station?

$v = 3 \times 10^8$ m/s

and $f = 198 \times 10^3$ Hz

$$\lambda = \frac{v}{f} = \frac{3 \times 10^8}{198 \times 10^3} = 1520 \text{ m}$$

Example

High frequency sound called ultrasound is transmitted through tissue at a speed of 1500 m/s and a wavelength of 0.03 m. Calculate the frequency of this ultrasound.

Solution

$v = 1500$ m/s

$\lambda = 0.03$ m

$$f = \frac{v}{\lambda} = \frac{1500}{0.03} = 50\,000 \text{ Hz}$$

Did you know?

The lowest frequency that humans can hear is 20 Hz and the highest is 20 000 Hz. Ultrasounds are frequencies of sound above the range of human hearing, that is greater than 20 000 Hz.

These frequencies can be used to detect shoals of fish at sea by measuring the time for the waves to reflect from the bottom of the seabed. If the time changes then something is in between the seabed and the boat. Used in this way it is called SONAR (Sound Navigation and Ranging) (Figure 9.9).

Other uses of ultrasound are in medicine where the waves are used in a similar way to check the progress of the baby in the womb (Figure 9.10). Here ultrasound is used because it will not cause damage to the unborn child. Kidney stones can also be shattered with pulses of ultrasound and opticians can use it to measure the depth of the eyeball (Figure 9.11).

Figure 9.9 *SONAR equipment*

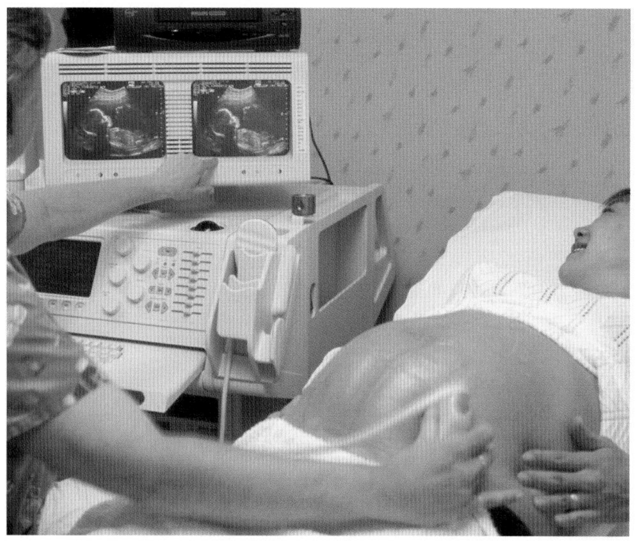

Figure 9.10 *Ultrasound scan of an unborn baby*

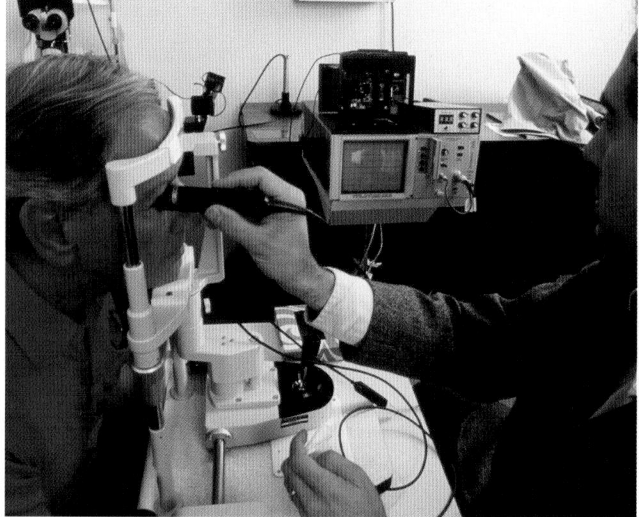

Figure 9.11 *Ultrasound scan by an optician*

Electromagnetic spectrum

The waves listed below travel at a speed of 3×10^8 m/s (300 000 000 m/s) in air. Each radiation has a different wavelength and frequency and will need a specific detector. Information about the different types of radiation is given in the table opposite.

	Radiation	Detector	Uses
Smallest wavelength and highest frequency	Gamma rays (from radioactive substances such as cobalt 60 or certain rocks or power stations)	Photographic film or Geiger counter	Kill cancer cells or bacteria in instruments
	X-rays (from X-ray tubes)	Photographic film	Produce pictures of bones inside the body
	Ultraviolet (Sun or certain lamps)	Skin or film which is said to fluoresce	Suntan
	Visible light (Sun)	Eye or photographic film	
	Infrared (lamps or very hot objects)	Photo transistor	Grills and toasters Remote controls
	Microwaves (cookers)	Aerial and receiver	Mobile phones and cooking
Largest wavelength and lowest frequency	TV and radio (transmitters)	Aerial and receiver	Transmit TV and radio signals

As the wavelength increases the frequency decreases but the product of frequency and wavelength is a constant, namely the speed of light.

Did you know?

Bluetooth devices

These are wireless devices that connect various devices together without cables, such as mobile phones and an earpiece and microphone combined in a headset. As well as mobile phones, computers, digital cameras and MP3 players all use this technology (Figure 9.12).

Unlike infra red devices such as remote controls, which must have a clear line of sight, bluetooth devices can operate as long as they are both switched on. The operating frequencies are in the range of 2.8 G Hz. A GHz is 1 000 000 000 Hz, or 10^9 Hz.

Figure 9.12 *A Bluetooth mobile phone headset*

Physics facts and key equations for waves

- A wave transfers energy.
- Radio, television signals and light are transmitted through air at 300 000 000 m/s.
- To calculate the various features of waves use $v = f \times \lambda$ or $v = \frac{d}{t}$.

- In a transverse wave the particles of the medium vibrate at right angles to the direction of the wave, an example being light.
- In a longitudinal wave the particles of the medium vibrate parallel to the wave direction, an example being a sound wave.
- The electromagnetic spectrum, from short to long wavelengths, is in this order: gamma rays, X-rays, ultraviolet, visible light, infrared, microwaves, TV and radio waves.

Questions

1 In an experiment to measure the speed of sound, a timer records how long a sound wave takes to travel between two microphones. The information recorded is:

distance between microphones = 3 m
time taken = 0.01 s

Calculate the speed of sound obtained from these results.

2 John is up a mountain range and shouts to his friend Heather. He hears the echo of his voice from a nearby mountain. The echo takes 6.6 s to return to him after he shouts. If the speed of sound is 330 m/s calculate the distance from John to the nearby mountain.

Figure 9.13

3 A patient is undergoing an ultrasound scan. A pulse of ultrasound is sent into the patient's body and is reflected from a bone back to a receiver. The time between the transmission and reception of the pulse is 0.00004 s. The speed of ultrasound in human tissue is 1500 m/s. Calculate the distance from the ultrasound transmitter to the bone.

4 When a water wave is viewed from the shore there appears to be an upward and downward movement of the waves. What is actually transferred by the wave?

5 a) For the wave shown in Figure 9.14 calculate:
(i) the amplitude of the wave
(ii) the wavelength of the wave.
b) 125 waves pass in 10 s. What is the frequency of the waves?
c) Calculate the wave speed.

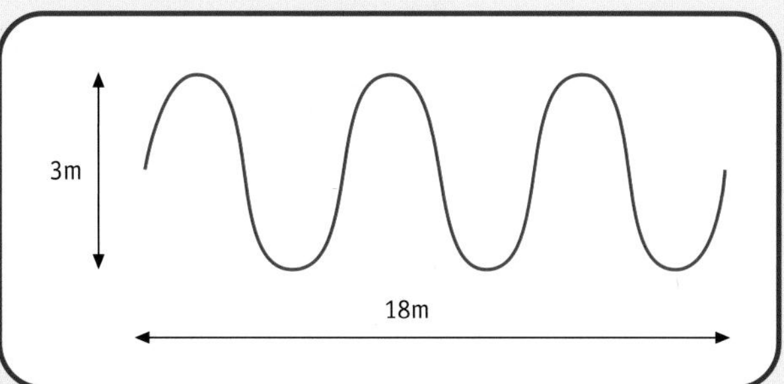

Figure 9.14

6 Copy and complete the table below.

Frequency (Hz)	Wavelength (m)	Speed (m/s)
0.3	6	
200	1.7	
1.5×10^{14}		3×10^{8}
250		1500
	1.5	200
	6×10^{-7}	3×10^{8}

7 A water wave has a wavelength of 4 m and a frequency of 5 Hz. Calculate the wave speed.
8 A wave has a speed of 1500 m/s and a frequency of 2 MHz. Calculate the wavelength of the wave.
9 A wave has a speed of 600 m/s and a wavelength of 0.025 m. Calculate the frequency of the wave.
10 State the speed at which radio and TV waves are transmitted.
11 a) Explain what is meant by a transverse wave and
b) give an example of a transverse wave.
12 Explain why a sound wave is a longitudinal wave rather than a transverse one.
13 Put the following waves in the correct order going from high frequency to low frequency.
gamma rays, TV, ultraviolet, visible light, microwaves
14 Which waves can be detected by the following instruments?
a) phototransistor
b) skin
c) photographic film.

10 Reflection

At the end of this chapter you should be able to...

1 State that light can be reflected.
2 Use correctly in context the terms angle of incidence, angle of reflection and normal when a ray of light is reflected from a plane mirror.
3 State the principle of reversibility of a ray path.
4 Explain the action of curved reflectors on certain received signals.
5 Explain the action of curved reflectors on certain transmitted signals.
6 Describe an application of curved reflectors used in telecommunication.
7 Explain, with the aid of a diagram, what is meant by total internal reflection.
8 Explain, with the aid of a diagram, what is meant by the 'critical angle'.
9 Describe the principle of operation of an optical fibre transmission system.

Law of reflection

A ray of light is shone from a ray box at a mirror. The angle of the ray is measured from a line drawn at right angles to the surface called a normal. This is the angle of incidence.

The angle of the reflected ray is measured. This is the angle of reflection. This can be repeated for other angles of incidence.

It is found that the angle of incidence equals the angle of reflection (Figure 10.1).

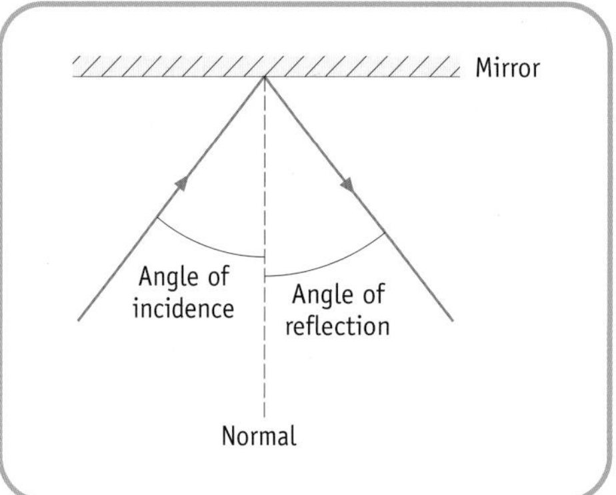

Figure 10.1 *Reflection of light*

This can be shown on a smooth pool of water where the scenery is easily reflected. If the water is disturbed and a rough surface is created then the reflected image is distorted (Figure 10.2 a and b).

Figure 10.2 *a) Clear reflection* *b) Distorted reflection*

If writing is looked at in a mirror then it is reversed, if a word is written backwards it will appear correctly in a mirror. This is often used in emergency vehicles since a driver looking in the rear view mirror will see the writing the correct way round and know that such a vehicle is behind the car (Figure 10.3).

Figure 10.3 *An emergency vehicle*

Reversibility

When a ray of light is shone back along the direction of one of the reflected rays, the ray is reflected back along the direction of the corresponding incident ray (Figure 10.4).

This is called the reversibility of the rays of light.

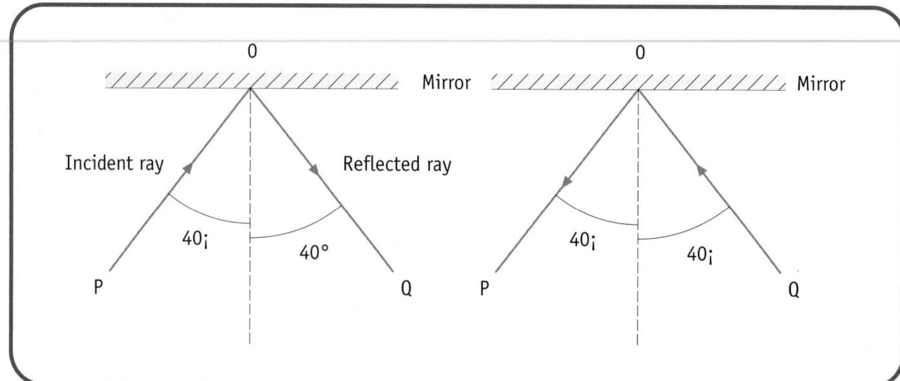

Figure 10.4 *Reversibility of rays of light*

Total internal reflection

A ray of light is shone into the semicircular glass block (Figure 10.5).

In diagram (a) most of the light emerges though the flat face of the block. As the angle of incidence is increased then the angle of the light in the air increases. In diagram (b) at a certain angle of incidence no light will pass from the glass through the flat face of the block. The direction of the light is along the surface of the glass block. When this occurs the angle of incidence is called

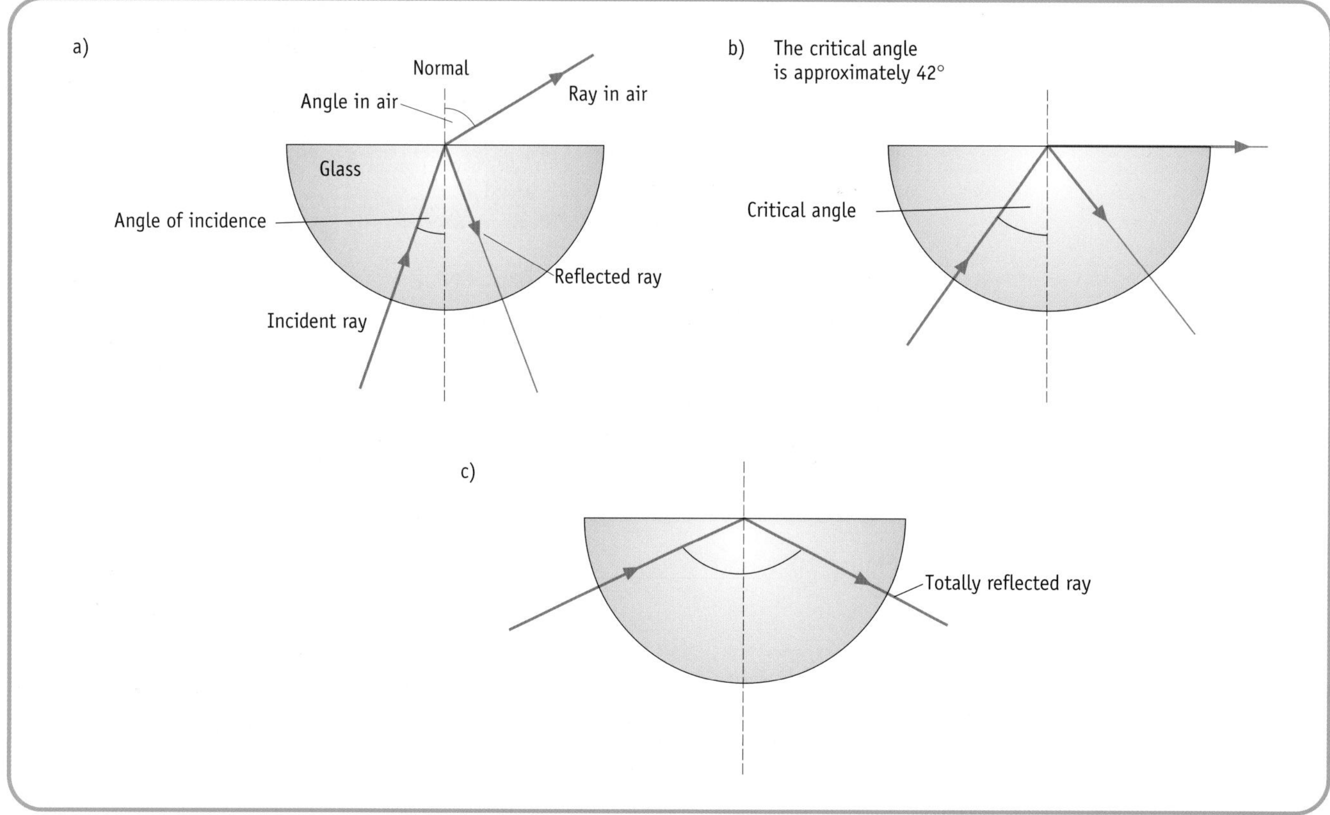

Figure 10.5 *Total internal reflection*

the critical angle. When the angle of incidence is increased further all of the light is reflected back into the glass. This is called total internal reflection. The critical angle for glass is about 42°. Beyond this angle the effect is total internal reflection and all light is reflected from the flat face of the block.

Did you know?

A periscope can be used for seeing above a certain height, most often in submarines but also at major events such as golf matches. The way to do this easily is to use two mirrors as shown in Figure 10.6a. When used in a submarine the silvering on the mirrors will peel away when exposed to the conditions at sea. The solution is to use two prisms as shown Figure 10.6b. The light enters the prism though one face at strikes the other face at 45°. This angle is greater than the critical angle and the light is totally internally reflected and after passing through the third face of the prism will repeat the process at the second prism.

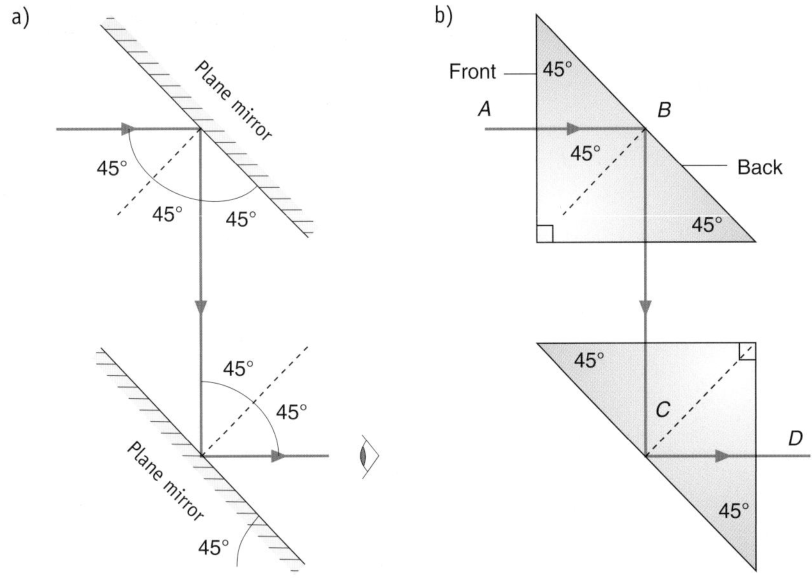

Figure 10.6 *Reflection from two prisms to make a periscope*

Optical fibres

Optical fibres are thin flexible glass threads about one eighth of a millimetre in diameter. Most of the glass thread is an outer layer of glass called cladding glass. In the middle of the thread is a different glass which is less than one hundredth of a millimetre in thickness. The fibres are made of extremely pure glass to cut down light loss and have a protective surface coating which reflects the light, keeping it inside the fibre.

An optical fibre is formed from glass so pure that a block 36 km thick would be as clear as an ordinary window pane. This means that we could see down to the bottom of the sea! Each fibre is no thicker than a strand of hair (Figures 10.7 and 10.8). The fibres use total internal reflection to send light signals from one end of the fibre to the other (Figure 10.9).

Figure 10.7 *Optical fibres*

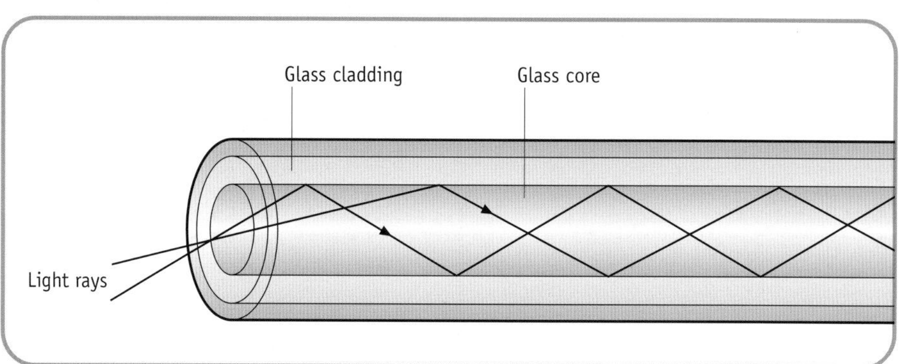

Figure 10.8 *Light rays in an optical fibre*

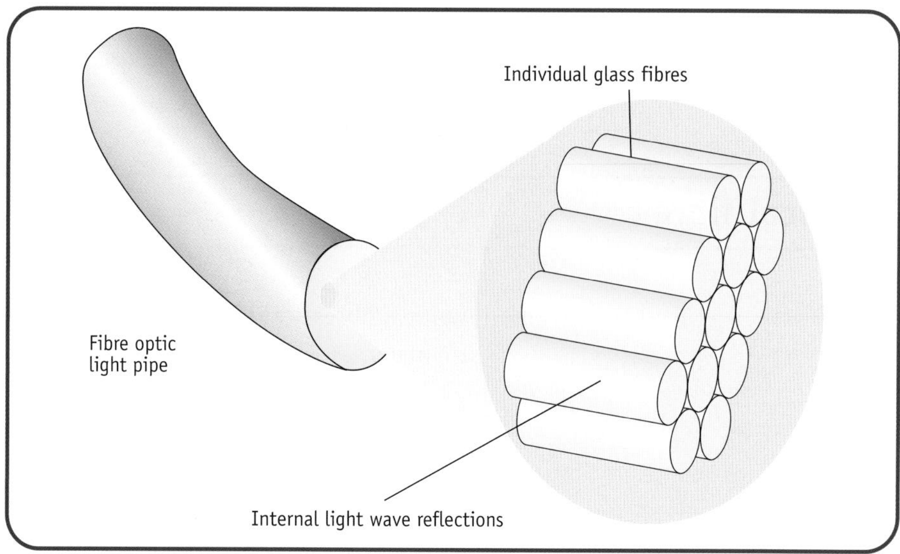

Figure 10.9 *Total internal reflection in fibre optics*

- Optical fibres are **lighter, carry more information** (up to 1000 telephone calls per fibre) and give better-quality communications than normal telephone wires.
- The signal that passes along the fibre is not electrical so it is less likely to be affected by other people's telephone calls or by other forms of electrical interference like mains hum.
- They are cheaper to make than copper since glass is mainly silica, which is cheaper than copper.
- They require fewer amplifiers called repeaters to transmit the signals over long distances.
- The disadvantage is that it is more difficult to join fibres together than copper wires.
- The speed of light in an optical fibre is 200 000 000 m/s.

Many modern telecommunication systems use optical fibres instead of copper wires. One single hair-like fibre can carry all the information needed to bring telephone messages, cable TV, videotext and computer services into your home.

The fibres operate by the light being totally internally reflected down the fibre, since no light can leave the outside of the fibre due to the angle of incidence being greater than the critical angle. It has been estimated that one optical fibre cable could take all the telephone calls being used at once in the world. This could create very large bills.

Transmission and detection

To send speech information along glass fibres it is first necessary to change sound signals into suitable pulses of electrical energy. The microphone (transmitter) in a telephone hand-set and some microelectronics do this. These pulses of electricity control a small laser (a narrow, very powerful beam of light), which then produces pulses of light that are transmitted through the optical fibre (Figure 10.10).

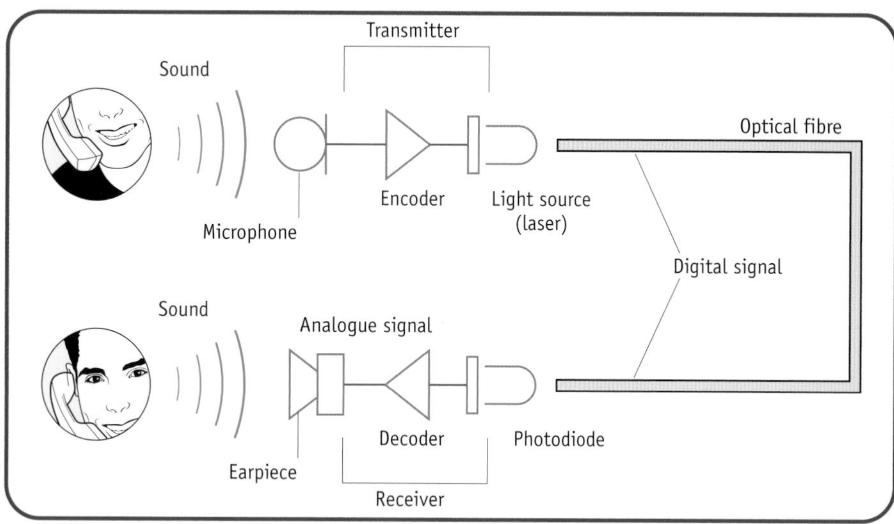

Figure 10.10 *Use of fibre optics in a simple system*

At the receiving end the light pulses are changed back into pulses of electricity by a small device called a photodiode. The electrical signal is fed to the earpiece (receiver), which then reproduces the original sound. In 1926, John Logie Baird invented a television system which used optical fibres, but it was not until some 40 years later that Charles Kao and George Hockham suggested that optical fibres might replace copper wires for telecommunications. In 1977 the world's first optical fibre telephone system became operational in America and in 1978 they were used in a town in Manitoba, Canada to carry telephone, television, radio and computer information.

All of Britain's major cities are now linked by the major fibre link trunk system (Figure 10.11). A transatlantic optical fibre link is now available.

Figure 10.11 *Installation of fibre optics*

Modern optical fibres transmit light signals with very little signal loss and can be used over distances of about 100 km without amplification. With conventional copper cables there is so much loss (or attenuation) of the signal that repeater amplifiers have to be installed every 4 km.

Did you know?

Diamonds are forever

Diamonds are an example of the use of total internal reflection to produce the sparkling effects. The critical angle for diamond is 24°. When the diamond is removed from the ground it is examined carefully so that when it is cut the maximum number of reflections can be achieved. by the person viewing it. Some of the light is reflected from the outer faces of the diamond but other rays are totally internally reflected and emerge from the inner surfaces. The effect of reflections gives a diamond its sparkle, called its 'fire'. A perfectly cut diamond will appear brilliant when viewed from the front and dark and dull when viewed from the rear since very little light passes through the diamond (Figure 10.12).

Figure 10.12 *Raw diamonds ready to be cut*

Dish aerials and satellites

Nowadays we can send and receive both television and telephone signals from nearly any part of the world. To do this we need a transmitting aerial dish and a receiving dish. These then send and receive signals from a satellite orbiting the Earth. These dish aerials use the physics of reflection from curved surfaces.

The dishes are curved so that parallel rays from a distant object are brought to a point called a focus. If the rays are radio waves then an aerial can be placed at this point to receive the waves.This is called a concave mirror and can be used for cosmetic or shaving mirrors. You can show the effect by holding the inside of a shiny spoon and if held far away an upside down image of the object will be seen (Figure 10.13).

Figure 10.13 *Reflection in a spoon*

Dish aerials

If a transmitting aerial is placed at the focus of a curved reflector (or dish), the reflected signal from the dish has the shape of a narrow beam (just like the light from a torch). This allows a strong (concentrated) signal to be sent in a particular direction from the transmitting dish aerial (Figure 10.14).

Figure 10.14 *A transmitting dish aerial*

Using dish aerials (curved reflectors)

When radio broadcasting began in 1920 the medium-frequency (MF) radio signals then used could travel about 1600 km. Soon after, high-frequency (HF) radio bands were discovered and these were used for worldwide communication.

Very high frequencies (VHF) soon followed, but these were unable to travel round the curvature of the Earth. Today VHF is used mainly for mobile communication, e.g. between aircraft and ground stations. As more and more information was transmitted the frequencies mentioned above became overcrowded and even higher frequencies had to be found. This led to the use of microwaves, whose frequency is about 10^9 Hz (1 GHz). Microwaves were useful since large amounts of information could be sent but little power was required. It was also found that they are easily focused using curved dish aerials (Figure 10.15).

Receiving dishes gather in most of the signal and reflect it to one point called the focus. The receiving aerial is placed at the focus to receive the strongest signal.

Microwaves are unable to diffract (bend) round obstacles because of their small wavelength. This means that the transmitting and receiving dish aerials must be in line of sight and are often located on towers. Since microwaves have a fairly short range a series of repeater (relay) stations are needed every 40 km. The incoming microwave signal is collected by the receiving dish. This signal is then reflected to the receiving aerial which is placed at the focus of the receiving dish. After being amplified (made bigger) the microwave signal

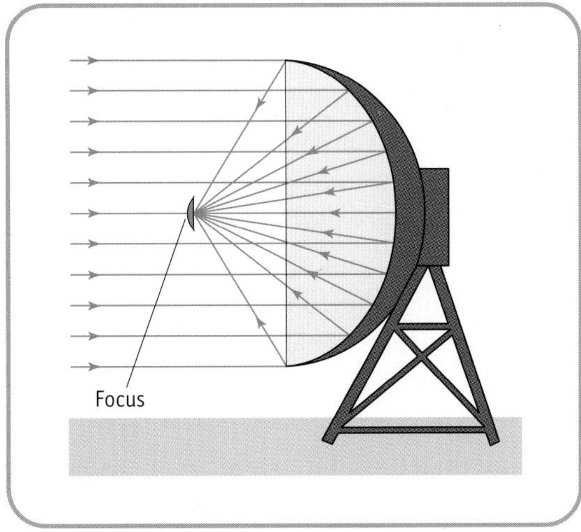

Figure 10.15 *A receiving dish aerial*

are passed to the transmitting aerial. This aerial is placed at the focus of the transmitting dish. The transmitted microwave is then reflected by the transmitting dish to produce a parallel narrow beam which is sent to the next repeater station. Using microwaves for round-the-world communication would require several hundred repeater (relay) stations at ground level. However, only three repeater stations in the sky, which are satellites, can cover all the earth if they are in the correct positions.

Satellites

A satellite has a very sensitive receiver as microwaves have to travel about 36 000 km to reach it. The signal, when received and focused, is amplified and transmitted back to Earth. Satellites have several different parts to them but

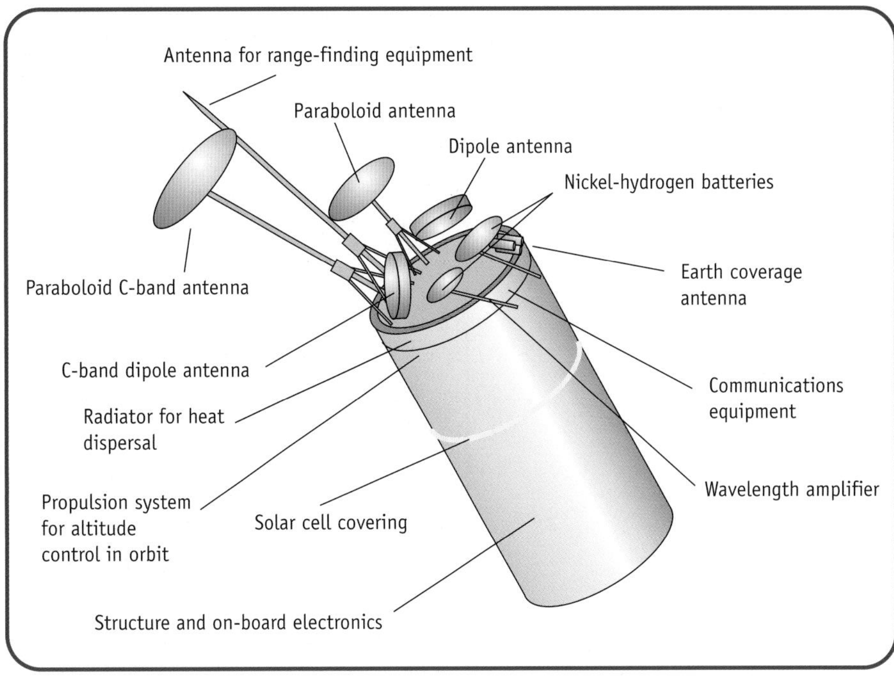

Figure 10.16 *Intelsat satellite*

they will all have two curved dish aerials, one for receiving signals from the Earth station and another to transmit them again after amplification. The received and transmitted signals are sent at different frequencies so that they do not interfere with each other. (Figure 10.16). The satellites are used not just for telecommunications but for weather information, surveillance and navigation both for boats, aircraft and now for cars. Several satellites are at dfferent heights although only three satellites were predicted to be needed to cover the whole Earth. Very large dish aerials on the Earth are required to pick up the weak signals coming from satellites and also to accurately send signals to them (Figure 10.17). The dish to receive the signals is the same shape as the transmitting one but the signals are received in a parallel beam and the rays are then brought to a focus. The diagram is the same but the direction of the arrows is reversed.

Figure 10.17 *Satellite dish*

Physics facts and key equations for reflection

- An optical fibre is a thin piece of glass along which light can be totally internally reflected.
- Modern television and telephone systems use electrical cables and optical fibres.
- When light is reflected from a plane mirror the angle of incidence = the angle of reflection.
- A curved reflector will bring the parallel rays of received signals to a focus, but reflects transmitted signals coming from the focus of the reflector into a parallel beam.
- Curved reflectors are used in satellite transmission and reception.
- Total internal reflection occurs when a ray of light does not leave the face of a piece of glass but reflects from that face back into the glass.
- The critical angle is the angle at which the ray inside the glass block passes along the surface of the glass when it emerges.
- An optical fibre transmission system changes electrical signals into light, sends the light along a fibre system and then changes the light back to electrical signals at the receiving system.

Questions

1 a) What is meant by an optical fibre?
   b) The diagram shows part of an optical fibre. Copy and complete the diagram to show the passage of light down the optical fibre.

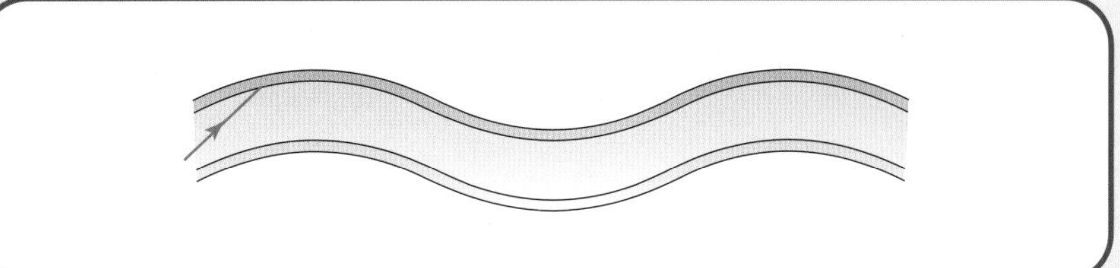

Figure 10.18

2 When light is shone onto a plane mirror it is reflected.
   a) Copy and complete the diagram to show what happens to the ray of light.
   b) Label your diagram to show the angle of incidence, angle of reflection and the normal.

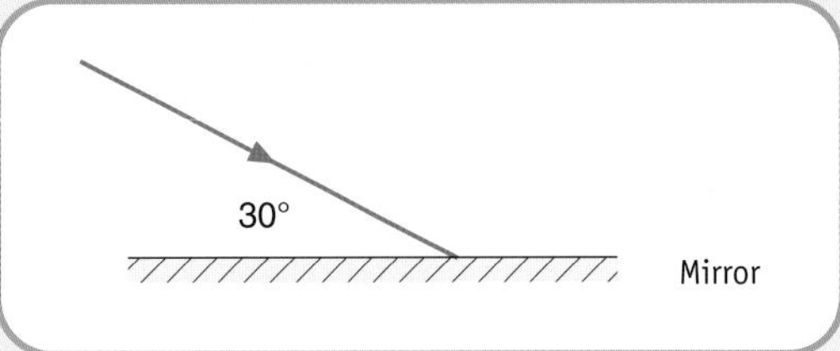

Figure 10.19

   c) For the diagram shown what is the value of the angle of reflection?

3 A local TV station is transmitting radio signals from a curved transmitter as shown.
   a) What happens to the signals as they leave the transmitting aerial?
   b) What name is given to the point where the transmitting aerial is placed?

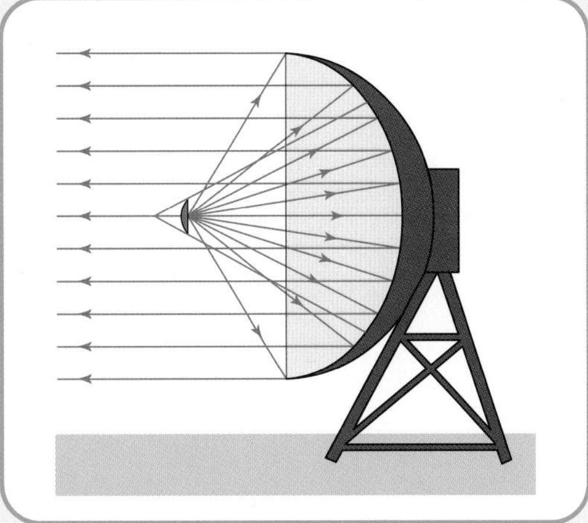

Figure 10.20

4 Satellite systems have two curved reflectors rather than just one. Use a diagram to explain the function of each reflector.

5 The glass prism shown below (Figure 10.21) is used in bike reflectors. The critical angle for the glass is 42°. Copy and complete the diagram to show the passage of the ray of light in the prism.

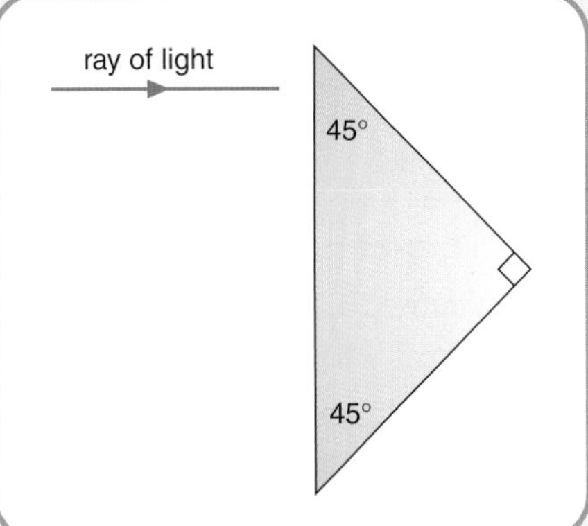

Figure 10.21

6 Modern communication systems use fibre optics to transmit information

a) What type of signal is sent along the optical fibre?

b) Describe how this signal is sent along the optical fibre.

c) State two advantages of using optical fibres.

Refraction

At the end of this chapter you should be able to...

1 State what is meant by refraction of light.
2 Draw diagrams to show the change of direction as light passes from air to glass and glass to air.
3 Use the correct terms angle of incidence, angle of refraction and normal.
4 Describe the shape of converging and diverging lenses.
5 Describe the effect of converging and diverging lenses on parallel rays of light.
6 Draw a ray diagram to show how a converging lens forms the image of an object placed at a distance of:
 a) more than two focal lengths
 b) between one and two focal lengths
 c) less than one focal length in front of the lens.
7 Carry out calculations involving the relationship between power and focal length of a lens.
8 State the meaning of long and short sight.
9 Explain the use of lenses to correct long sight and short sight.

Refraction of light

In a given material (called a medium) light travels in a straight line. When the light travels from one material to another it may bend or change direction as it enters the new material. When light enters a new material its speed changes. This effect is called **refraction**.

Rays of light can travel through various objects. The paths of the rays passing through and leaving the objects can be drawn as shown in Figure 11.1. The dotted line drawn at right angles to the surface is called the normal.

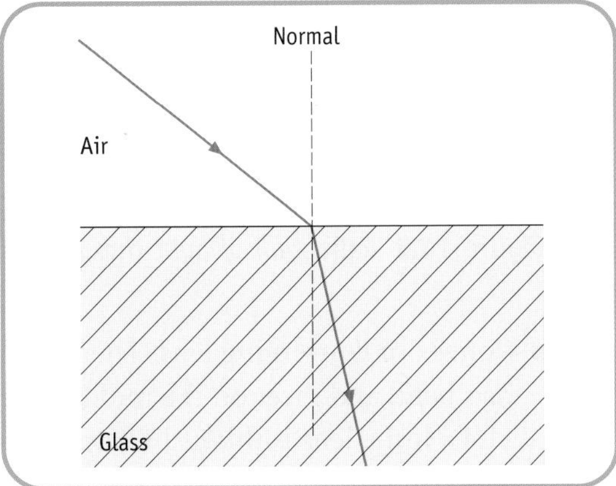

Figure 11.1 *Refraction of light*

- **Plane rectangular block.** When the incident ray travels parallel to the normal, there is no change in direction. When the incident ray is at an angle to a plane rectangular block, the ray coming from the block is parallel to the incident ray (Figure 11.2). The angle of the ray of light striking the block is called the **angle of incidence** and is **always greater** than the ray of light inside the block called **the angle of refraction.**
  Remember all angles are measured against the normal.

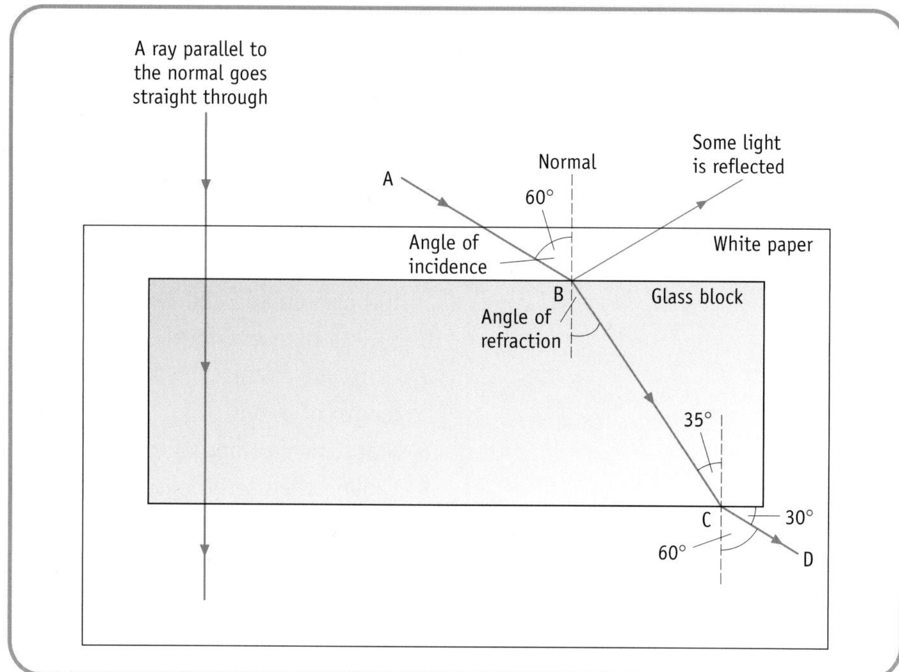

Figure 11.2 *Refraction of light through a rectangular block*

- **Triangular prism.** The ray bends towards the normal going into the prism and away coming out (Figure 11.3).

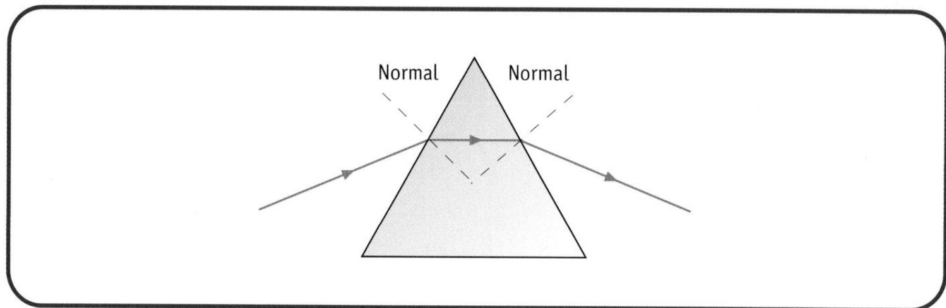

Figure 11.3 *Refraction of light through a triangular prism*

- **Convex or converging lens.** The middle ray goes straight on and the outer rays bend and meet on the middle line at a point called the focus (Figure 11.4). If the lens is thick, the same effect occurs but the focus is nearer the lens (Figure 11.5).

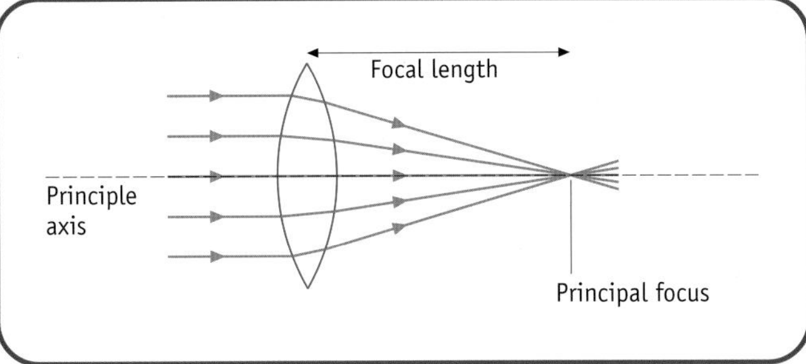

Figure 11.4 *A thin lens is a weak lens. It has a longer focal length than the strong lens*

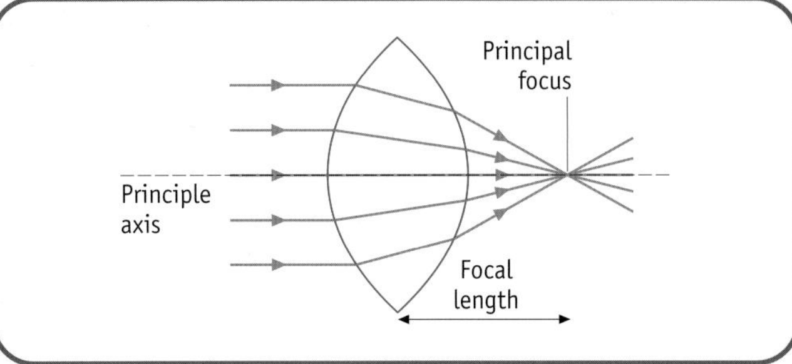

Figure 11.5 *A thick lens is a strong lens. It has a short focal length*

- **Concave or diverging lens.** The rays spread out and appear to come from a focus behind the lens (Figure 11.6).

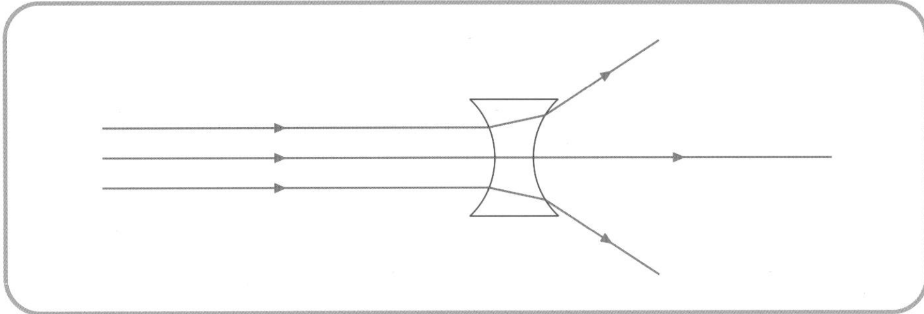

Figure 11.6 *A concave or diverging lens*

Did you know?

A common effect of refraction is the illusion of an object at a different depth in water than you think. In the diagram the rays of light from the fish bend away from the normal when leaving the water and enter your eye. The brain thinks the rays must have travelled in straight lines and we think the fish is closer to the surface. To the fish the world appears distorted due to refraction and reflection (Figure 11.7).

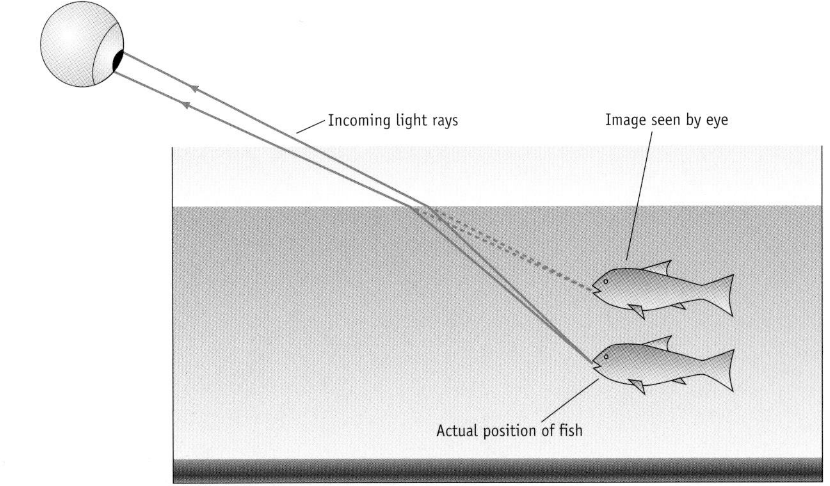

Figure 11.7 *Virtual image observed by refraction*

Focal length

Some lenses bend light more than others. This is due to their thickness and the amount of curving of the lens. One way to indicate the amount of refraction is to measure the focal length of the lens. A convex or converging lens can make rays of light come together to a point after they have passed through the lens. The point where the rays meet is called the focus. The position of the focus depends on where the rays come from. When the rays come from a distant object which is so far away that the rays are parallel, the focus is closer to the lens. In this case, the focus is called the principal focus. **The distance from the lens to the principal focus** is called the **focal length** and is measured in **metres.**

Ray diagrams and lenses

An object such as a light source can be placed at different distances from a convex lens. The effect of placing the source at different distances has different effects on the final image. It is possible to know the effect by drawing an accurate diagram.

In doing these diagrams certain rules apply:

- The lens is considered thin enough to be a line and refraction only takes place at this line.
- Thousands of rays come from the object but it is possible to consider only the minimum number of rays to complete the diagram.
- One ray is drawn from the top of the object parallel to the central line called the principal axis. This ray then passes through the principal focus.
- A second ray passes straight from the top of the object through the centre of the lens without changing direction.
- A ray can be drawn through the principal focus on the first side of the lens and then after refraction by the lens can be drawn parallel to the principal axis.
- Where the rays intersect is the top of the object (Figure 11.8).

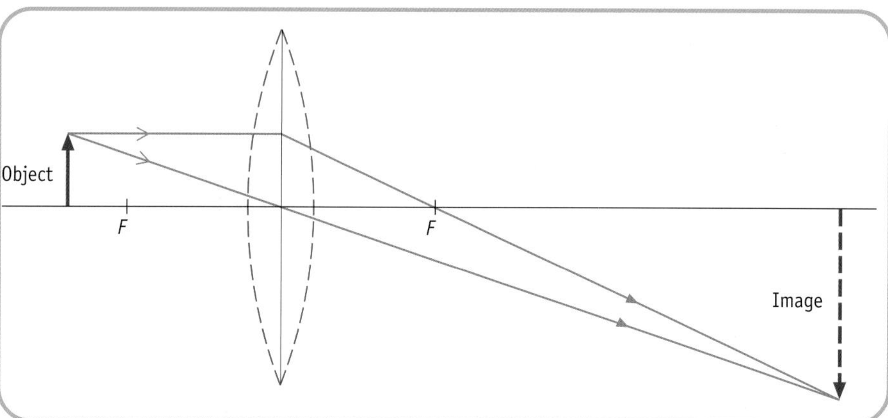

Figure 11.8 *Rays passing through a convex lens*

- The image can be described in three ways:
  a) It is real or virtual. A real image can actually be seen on a screen. A virtual one is one that appears to come from that point. It is an illusion of the eye and brain and cannot be caught on a screen.
  b) It is magnified or diminished. If you measure the size of the image and compare it to the size of the object then you can easily find the change in size. This is called the magnification of the lens.

  $$\text{Magnification} = \frac{\text{length of image}}{\text{length of object}}$$

  c) Upright or inverted. The image will either be the same way up as the object or it will be upside down.

Three cases are shown in Figure 11.9:

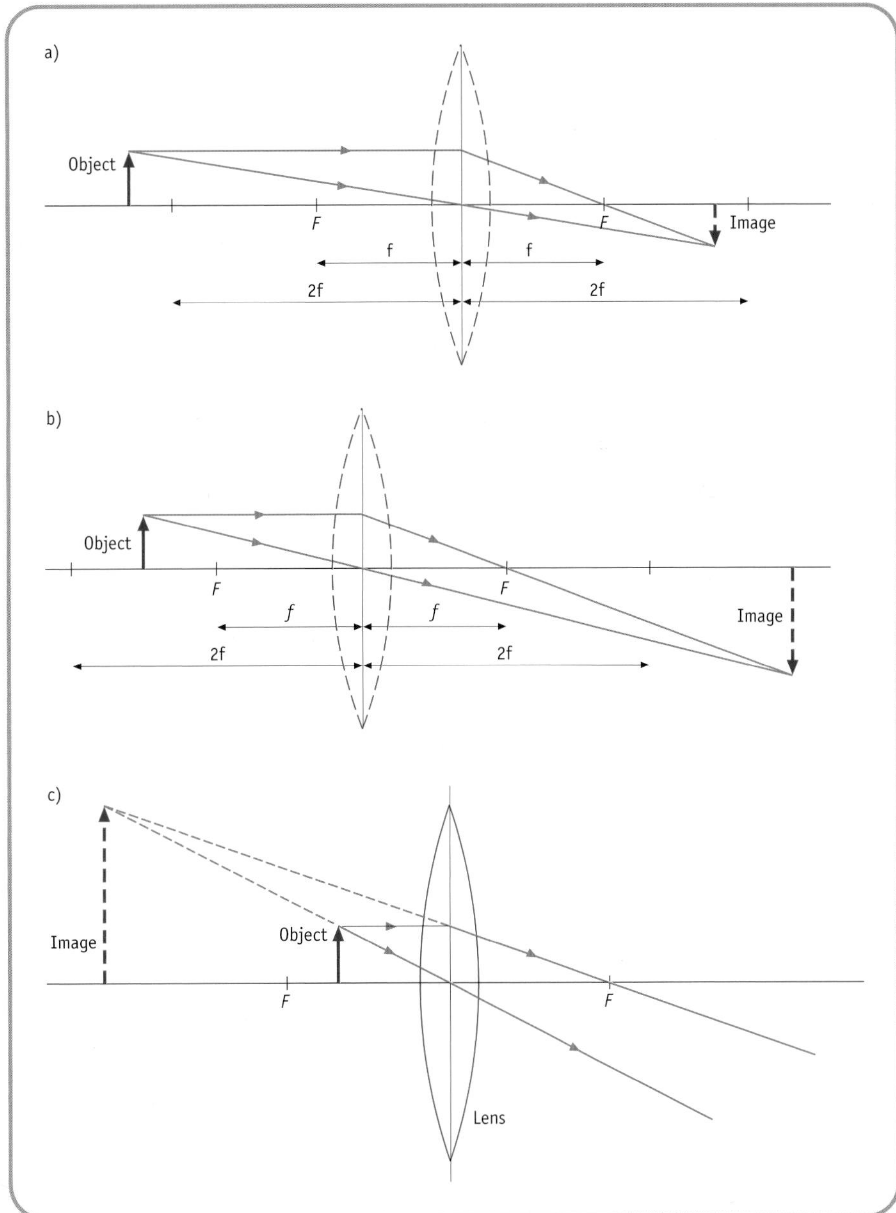

Figure 11.9 *Objects viewed through a convex lens*

a) Object **more than two focal lengths:**
the image is real, inverted and diminished. This is the normal image on the retina of the eye.

b) **Object between one and two focal lengths:**
the image is real, inverted and magnified. An overhead projector or a slide projector works in this way

c) **Object less than one focal length:**
Here the rays do not meet unless they are traced back. The image is virtual, upright and magnified.

The last example is how a magnifying glass works. The eye thinks the image is magnified but the image is virtual since it cannot be caught on a screen.

The eye

Lenses are often used to correct eyesight defects. It is important to understand how the eye operates. An outline of the eye is shown in Figure 11.10. The different parts of the eye and their functions are as follows:

- Light enters the front of the eye at the cornea. This is transparent and it is here that most of the refraction or bending of light occurs.
- The light enters a lens, which is a jelly-like substance, and more refraction occurs.
- The lens is held by fibres which act like muscles. These can change the shape of the lens from thick to thin.
- The light then passes through a gel-like substance which makes the light spread out.

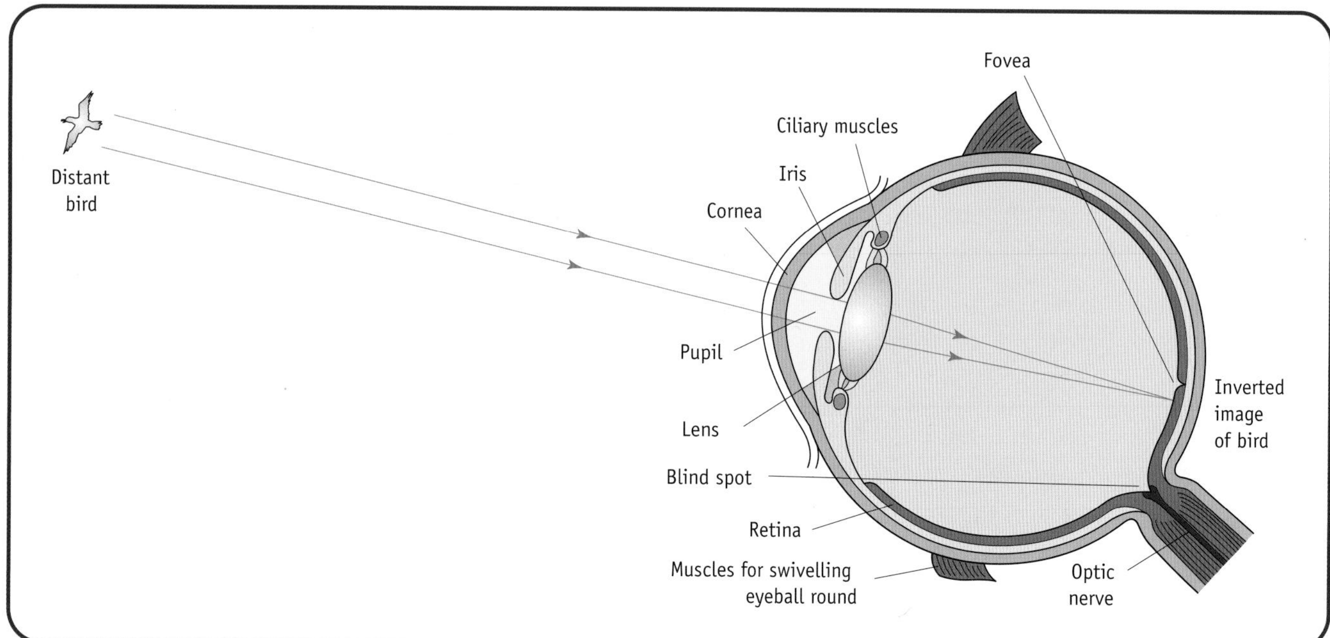

Figure 11.10 *Light passing through the eye*

- The light reaches the retina at the back of the eye.
- To see objects clearly, the light must be focused on the retina.

Electrical signals pass along the nerve fibres to the brain. The part of the retina where the nerve fibres leave the retina contains no light-sensitive cells and is therefore a blind spot on the retina.

The amount of refraction which takes place at the cornea does not change. However, in order to focus on near and on distant objects, an adjustable lens is needed. This is provided by the eye's lens. This lens is held by muscle-like fibres in the ciliary body which can change the shape of the lens from thick to thin. The lens is thin and can focus on distant objects. To view near objects the muscles change the lens shape to thick. When light enters the eye, the image formed on the retina is upside down. The brain learns to turn this image the 'right way up'.

Power of a lens

People who have severe eyesight defects may need stronger (more powerful) lenses to correct their eyesight defect than those who have only slight defects. (A more powerful lens is one which causes more refraction.) An optician must therefore have a range of lenses to suit different individuals' needs. The powers of these lenses could be indicated by giving their focal lengths: the most powerful lenses having the shortest focal lengths. Another way to indicate the amount of refraction caused by a lens is to calculate its power from the equation:

$$\text{power} = \frac{1}{\text{focal length}}$$

where the focal length is measured in metres and the power is given in dioptres (D).

Converging (convex) lenses have **positive** powers (e.g. +10 D, +17 D).

Diverging (concave) lenses have **negative** powers (e.g. −10 D, −17 D).

Example

A convex lens has a focal length of 100 mm. Find the power of the lens.

Solution

focal length = 0.1 m

$$\text{power} = \frac{1}{\text{focal length}} = \frac{1}{0.1} = 10\ \text{D}$$

Example

A lens has power of −2 D. Calculate its focal length.

Solution

Power = −2 D, which tells us that this a concave lens since there is a negative sign.

$$\text{Power} = \frac{1}{\text{focal length}}$$

$$2 = \frac{1}{\text{focal length}}$$

$$\text{focal length} = \frac{1}{2}$$

$$= 0.5 \text{ m}$$

Eyesight Defects

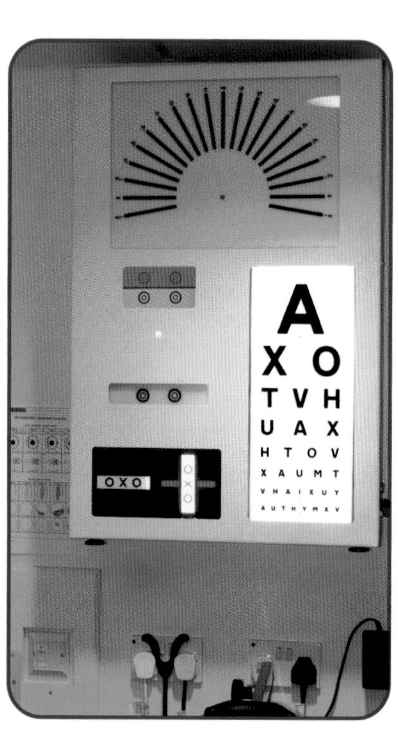

Figure 11.11 *Snellan chart*

There are a large range of eyesight defects which can cause difficulties with vision. The eye testing procedure involves the physical examination of the eye and a chart called a Snellen chart which you view at a distance of 6 m (20ft). (Figure 11.11). Most rooms are smaller than this so the use of a mirror is made and the chart is behind you and you look at the mirror image. Normal vision is said to be 20/20 when you can see a line of type of a certain size at a distance of 6 m. If you have an eyesight defect then you would have say 20/40 vision. This means that at 6 m you can only read a much larger size of type than the person with normal vision.

There are two sight defects which a large number of people will experience – namely long and short sight.

Long sight

A long-sighted person can see far-away objects clearly. Objects quite close to the eye appear blurred. The eye lens is bringing the rays to a focus beyond the retina. This may happen if the person's eyeball is shorter than normal from front to back. It may also be caused by ciliary muscles which cannot relax for the lens to be made fat enough.

A converging (convex) lens corrects this fault since it will increase the bending of the light rays before they enter the eye lens and so the light will be focused on the retina. The object is thus seen clearly (Figure 11.12).

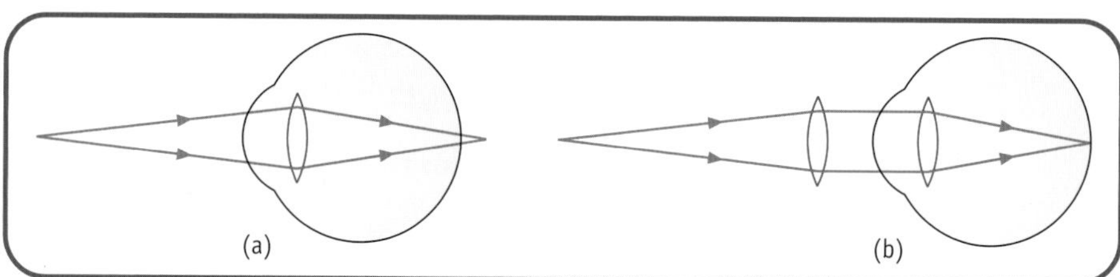

Figure 11.12 *a) A long-sighted eye cannot see near objects clearly b) A converging lens corrects long sight*

Short sight

A short-sighted person finds that distant objects are blurred but near objects are in focus. The eye lens is bringing light to a focus in front of the retina, which may happen due to the lens having a large curvature. The muscles cannot make the lens thin enough.

A diverging (concave) lens corrects this fault, since it will spread the light out more before it enters the eye lens, and so the light will be focused on the retina (Figure 11.13).

For revision purposes remember either short or long sight and its correcting lens and the other sight defect is the opposite.

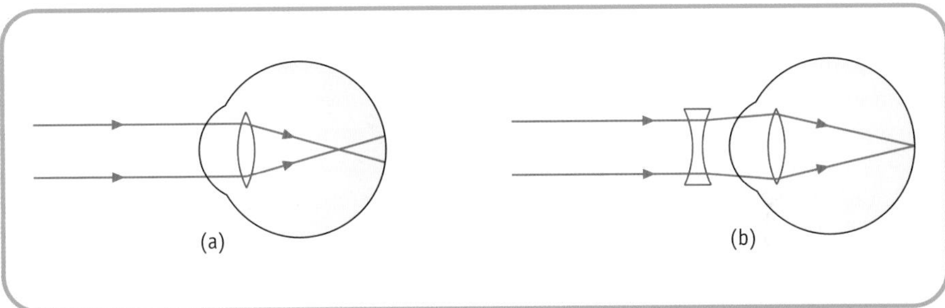

Figure 11.13 *a) A short-sighted eye cannot see distant objects clearly b) A diverging lens corrects short sight*

Did you know?

The first spectacle lenses were developed around about 1270 by joining two magnifying glasses together but there is a legend that the emperor Nero had an emerald cut as a lens to enable him to see more clearly.
Contact lenses were first suggested by Leonardo da Vinci in 1518, who noticed that he could see more clearly when he opened his eye under a bowl of water. Different lenses have been developed that allow oxygen to pass into the eye, which helps to prevent some eye diseases. There are lenses which can be thrown away after use each day and also ones which have a varying degree of curvature (Figure 11.14).

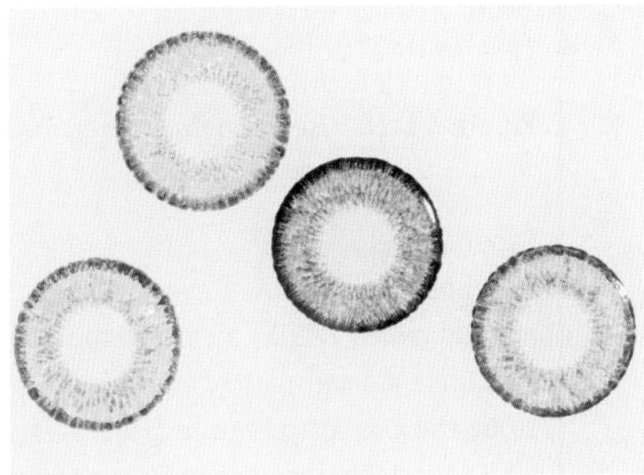

Figure 11.14 *Contact lenses*

The latest technique does not use any surgery but uses lenses at night. Lenses are placed in your eyes when you go to sleep and taken out when you wake up. During the night the lenses flatten your cornea so that you can see reasonably well for most of the day without contact or spectacle lenses. The technique is especially useful in situations where you might have your eyes watering, for example, playing sports on a very warm day or if you are a firefighter entering a smoke filled area (Figure 11.15).

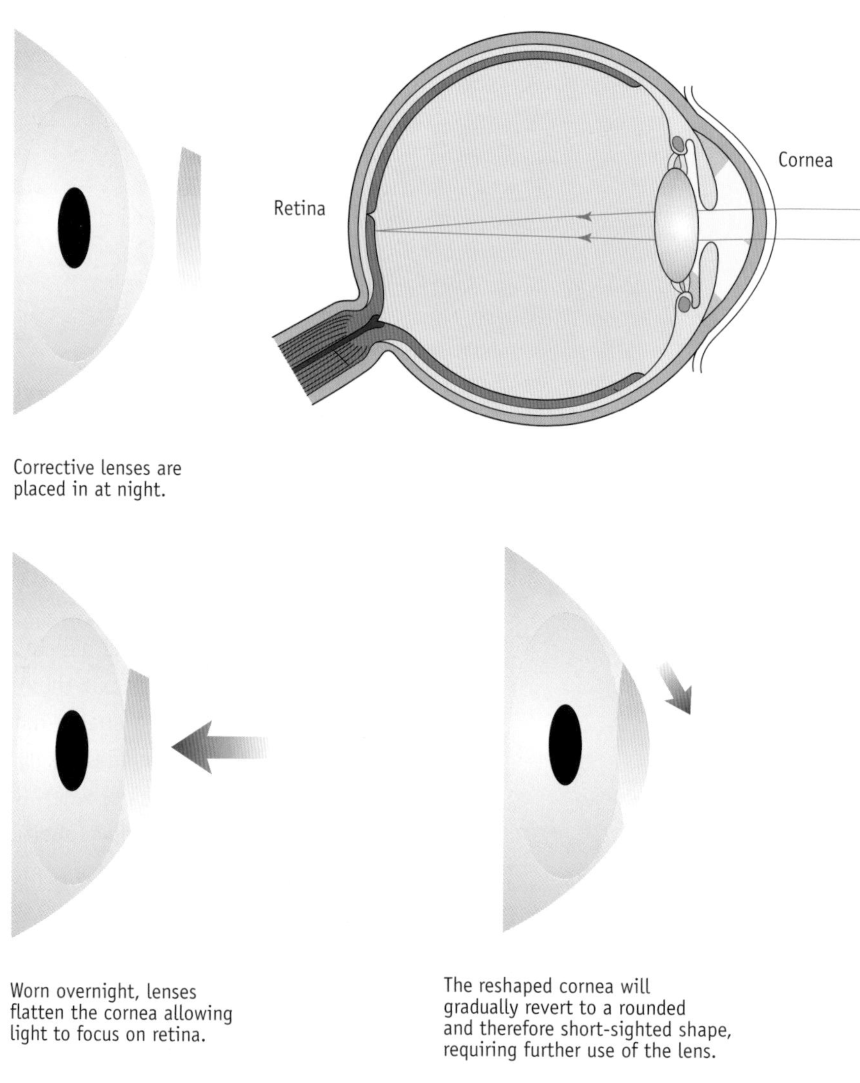

Figure 11.15 *Use of contact lenses overnight to correct vision*

Laser surgery

It is possible to correct eyesight defects with surgery. In the most common method called LASIK, a small flap is cut in the cornea and a laser is used to vapourise a small part of the thickness of the cornea. The cornea is reshaped to be the correct curvature (Figure 11.16).

While the technique seems very good if there are any problems then the removal of tissue cannot be reversed.

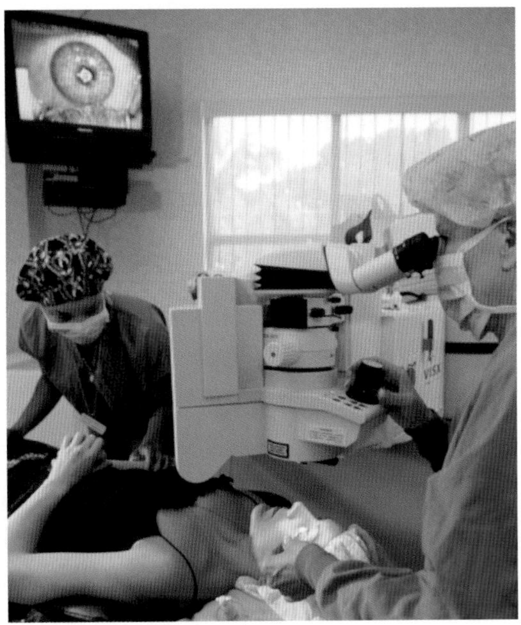

Figure 11.16 *A LASIK operation*

Physics facts and key equations for refraction

- Refraction of light occurs when light travels from one substance to another.
- When light travels from air to glass the ray bends towards the normal and the angle of incidence is greater than the angle of refraction.
- A converging lens brings parallel rays to a focus but a diverging lens spreads the rays out.
- For a converging lens, the image of an object placed at a distance of:
  a) more than two focal lengths is real, diminished and inverted
  b) between one and two focal lengths is real, magnified and inverted
  c) less than one focal length is virtual, magnified and upright.
- The power of a lens in dioptres is calculated as:

  $$\text{Power} = \frac{1}{\text{focal length}}$$

  where the focal length is in metres.
- Long sight is when distant objects are seen clearly but near objects are blurred. A converging lens will correct the problem.
- Short sight is when distant objects are blurred but near objects are clear. A diverging lens will correct the problem.

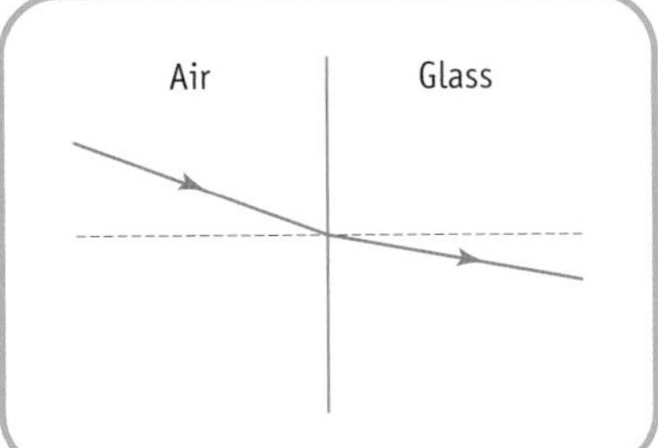

Figure 11.17

Questions

1 A ray of light passes from air to glass as shown in Figure 11.17.
  a) What is the name of this effect?
  b) Copy the diagram and label the angle of incidence and the angle of refraction.
  c) What can you state about the size of the angle of incidence compared to the angle of refraction?

2 Copy and complete the diagrams below to show how the rays of light pass through the blocks of glass. Show clearly the normal and label the angles of incidence and refraction.

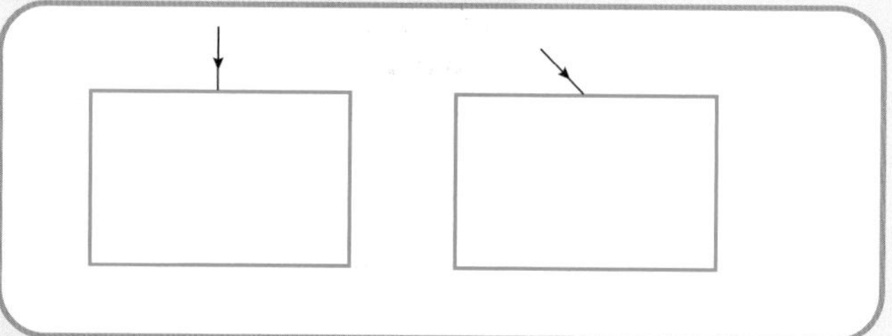

Figure 11.18

3 A convex lens is used to view a plant. The diagram below shows a plant which is placed at a distance of 250 mm in front of a convex lens. The focal length of the lens is 100 mm.

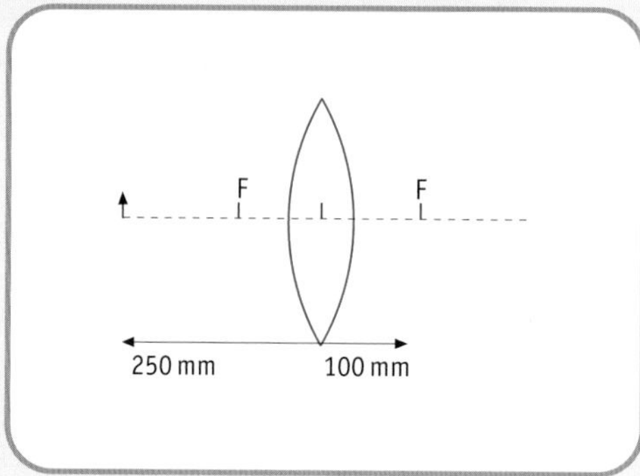

Figure 11.19

a) Draw a suitable diagram to scale to show the final image produced by the lens.
b) Measure the height of the final image. Calculate the change in size of the image compared to the object.

4 A postage stamp is viewed using a magnifying glass. As the magnifying glass is moved closer to the stamp the image changes from upside down to upright.
a) At what lens distance will this effect begin to occur?
b) The lens has a focal length of 50 mm and the stamp is placed 30 mm from the centre of the lens.
   (i) Using a suitable diagram show how the image of the stamp is produced.
   (ii) Calculate the magnification of the stamp.

5 A slide is placed at 150 mm from the centre of a lens in a slide projector. The projector lens has a focal length of 100 mm.
Draw a diagram to show how the image of the slide is produced.

6 An object is placed in front of different convex lenses.
For each situation:
- Draw a ray diagram to show how the image of the object is formed.

Measure and state the distance from the lens to the position of the image.

- State whether the image is real or virtual.
- State whether the image is larger or smaller than the object.

a) The object is placed 90 mm from a convex lens of focal length 40 mm

b) The object is placed 70 mm from a convex lens of focal length 50 mm

c) The object is placed 30 mm from a convex lens of focal length 450 mm

7 A prescription for a lens states that a correcting lens of + 2.5 D is required.

a) What type of sight defect is the person suffering from? Explain your answer.

b) Calculate the focal length of the correcting lens.

8 Copy and complete the table for the power and focal length of the lens and state if the lens is convex or concave.

Focal length (mm)	Power (dioptres)	Type of lens
100		Convex
250		Concave
	4	
	−5	
300		Concave
400		Convex

9 A student can read this book without the aid of glasses. He needs spectacles to see the number on an approaching bus.

a) Name the sight defect that the student is experiencing.

b) What type of lens is needed to correct this defect?

c) The focal length of the lens in the student's glasses is 0.6 m. Calculate the power of the lens.

10 Three shapes of contact lenses are shown below (Figure 11.20).

a) State and explain which lens or lenses could be used to correct

i) long sight

ii) short sight.

b) What advantage do contact lenses have over conventional spectacle lenses?

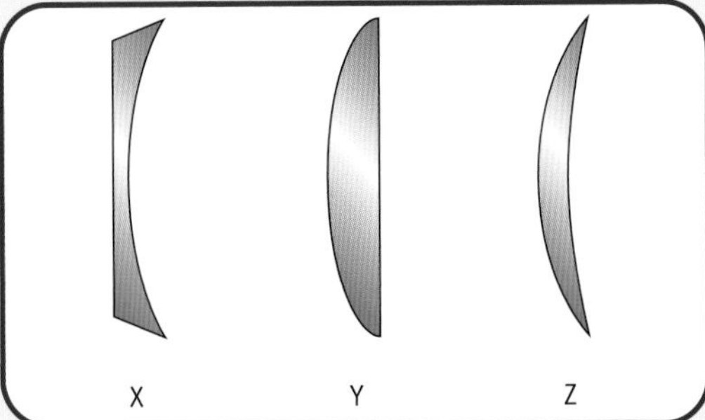

Figure 11.20

Exam Questions

1 During an investigation of lenses, a small lamp of height 10 mm is placed at a distance of 300 mm from a convex lens of focal length 200 mm.

a) (i) By using a suitable scale diagram find where the image is located.

(ii) State whether the image is real or virtual and whether it is magnified or diminished.

b) This same lens can be used by someone with a sight defect.

(i) Which sight defect can be corrected by using a convex lens?

(ii) Calculate the power of this lens.

2 A radio controlled toy operates on a frequency of 27 MHz.

a) (i) Explain what is meant by 'frequency of 27 MHz'.

(ii) Calculate the wavelength of the transmitted wave.

b) Modern communication systems use fibre optics as a method of sending signals.

(i) What is meant by a fibre optic sytem?

(ii) An optical fibre can normally send a signal a distance of 40 km before amplification is required. Calculate the time required to send this signal a distance of 40 km down the fibre. The speed of light in the fibre is 2×10^8 m/s.

3 Ultrasound is high frequency sound beyond the normal range of hearing. It can be used in medicine and industry. Fishing boats use ultrasound to detect shoals of fish under the water.

The frequency of the ultrasound used is 5 MHz. The speed of ultrasound in water is 1500 m/s.

a) Calculate the wavelength of this ultrasound.

b) The ultrasound is sent down to an apparent shoal of fish. The time between transmission and reception of the ultrasound is 0.24 s. Calculate the depth of the water where the shoal of fish are located.

4 The diagram below shows a ray of light entering a glass block from air (Figure E1).

a) What is the value of the angle of refraction?

b) What is the value of the angle of incidence?

c) Many people need spectacles and some people need two type of lenses in one frame. One for close detailed work and another for viewing distant objects.

The powers for the lenses are + 1.5 D and + 3.5 D. State and explain which lens is used for close detailed work.

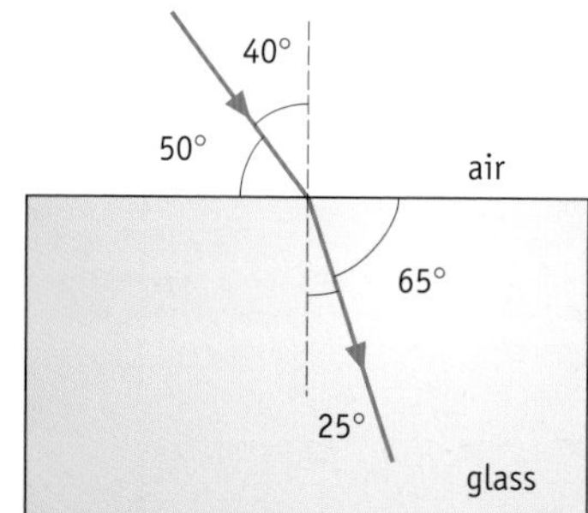

Figure E.1

Radioactivity

A low level of radiation exists naturally and is all around us, but scientists can produce higher levels for use in medicine, power supply and industry. This is a complex and potentially dangerous activity, and this unit explores the potential, issues and problems involved in using radioactivity in this way.

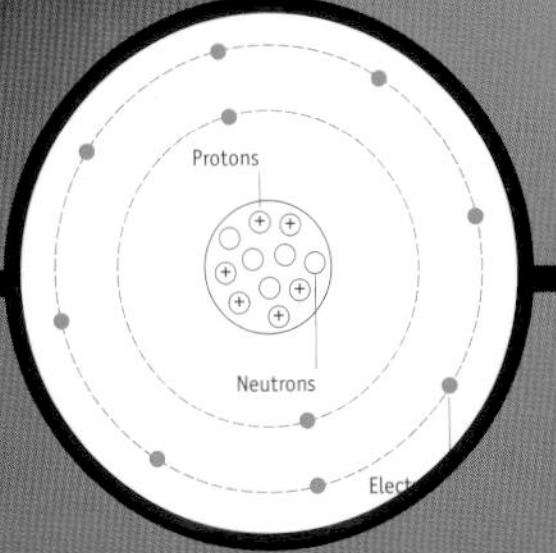

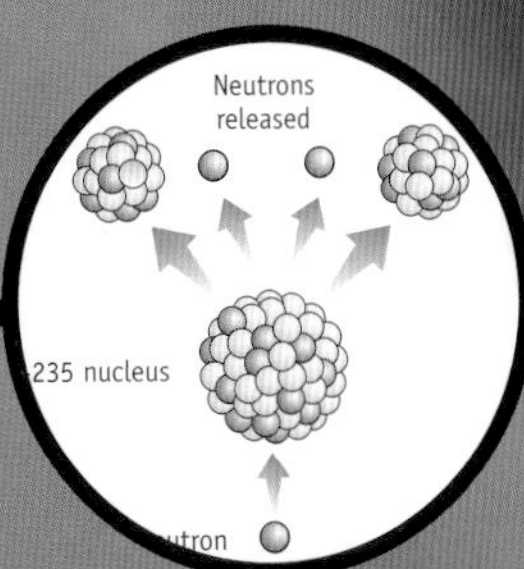

12 Ionising radiation

At the end of this chapter you should be able to...

1 Describe a simple model of the atom which includes protons, neutrons and electrons.
2 State what is meant by an alpha particle, beta particle and gamma radiation.
3 State that radiation energy may be absorbed in the medium through which it passes.
4 State the range and absorption of alpha, beta and gamma radiation.
5 Explain the term ionisation.
6 State that alpha particles produce much greater ionisation density than beta particles or gamma rays.
7 Describe how one of the effects of radiation is used in a detector of radiation.
8 State that radiation can kill living cells or change the nature of living cells.
9 Describe one medical use of radiation based on the fact that radiation can destroy cells.
10 Describe one use of radiation based on the fact that radiation is easy to detect.

Atoms are the smallest possible particles of the elements which make up everything around us. All atoms of the one element are identical to one another, but they are different from all other elements. This is because they are made up from different combinations of electrons, protons and neutrons. These are the three types of particles which make up an atom.

All atoms have a tiny central nucleus, which has a positive charge. We can imagine the negatively charged electrons to be circling around this nucleus. The nucleus contains the positive protons and the neutrons, which are uncharged (see Figure 12.1 opposite). The mass of the protons and the neutrons are almost identical but the electrons have a very small mass compared to protons or neutrons.

Different types of radiation

There are three types of radiation:

1 Alpha (α) particles.
2 Beta (β) particles.
3 Gamma (γ) rays.

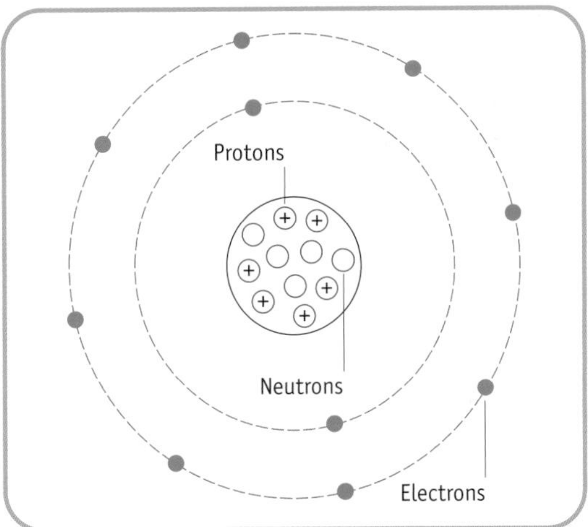

Figure 12.1 *Diagram of an atom*

These can be identified by what happens as they reach different materials.

Alpha radiation can be completely stopped by a few sheets of paper. It is a helium nucleus that is a helium atom which does not have the outer electrons.

Beta radiation can be absorbed by a sheet of aluminium.They are fast moving electrons.

Gamma radiation is reduced and may be completely absorbed by materials such as concrete, lead and other dense materials. The thicker the material the more radiation that will be absorbed. This use will be discussed later as a form of protection.

When the alpha or beta or gamma radiation passes through a material it loses energy by colliding with the atoms of the material. Eventually the radiations lose so much energy that they cannot get through (penetrate) the material and so are absorbed.

Information about the three different types of radiation and their natures are shown in the following table.

Type of radiation	Symbol	What is this radiation	Charge and absorption
Alpha	α	Helium nucleus, i.e. two protons and two neutrons	2+ Absorbed by paper or about 5 cm of air
Beta	β	Fast-moving electron	1− Absorbed by aluminium
Gamma	γ	Short-wavelength electromagnetic radiation	Uncharged Absorbed by lead

Ionisation

If an electron is added or removed from an atom, what is left is called an **ion**. The process is called ionisation. The removal of an electron creates a positive ion and if an electron is added then a negative ion is formed. It is the breaking up of an atom into positive and negative parts when alpha, beta or gamma radiation passes close to the atom.

The process of ionisation by an alpha particle is shown in Figure 12.2. In (a) the alpha particle is approaching the neutral atom and in (b) it has passed by having created an ion pair. This means that the alpha particle has caused the atom to lose an electron and form an ion.

Since an alpha particle moves much slower than beta or gamma radiation it will cause much more ionisation. The amount of ionisation produced in a certain volume is called the **ionisation density**.

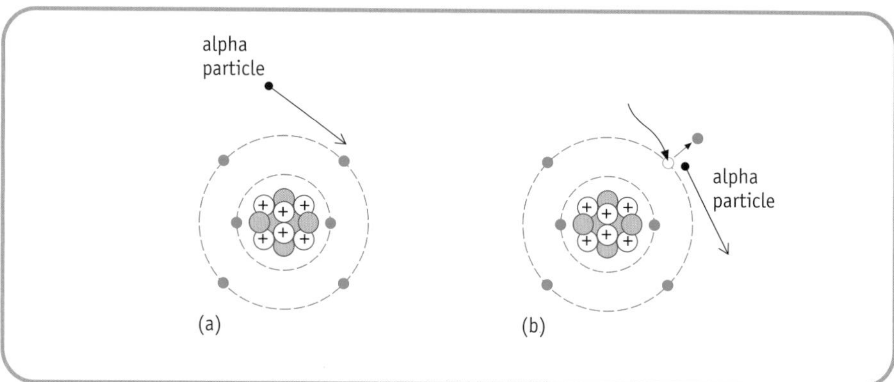

Figure 12.2 *a) Neutral carbon atom. b) Ionisation of carbon atom*

Effect on the human body

- Alpha radiation will produce ionisation in a short distance of body tissue. This type of radiation outside the body is absorbed by the skin and little damage will occur. Swallowing the radiation will produce large amounts of ionisation and would be dangerous. In a small volume with a large number of ions this is called a high ionisation density.
- Beta radiation will be absorbed in about 1 cm of tissue and any beta radiation outside the body will cause damage to that tissue but a small

amount can penetrate the body. If the radioactive source is put into the body then internal organs can be damaged.

- Gamma rays pass through the body and can damage tissue whether the source of radiation is inside or outside of the body. Gamma rays are able to pass from inside the body to outside the body. Alpha and beta radiation cannot escape from inside the body.

Sterilisation

Radiation can be used to kill cells, and can also be used to kill bacteria or germs. Previously, medical instruments such as glass syringes had to be sterilised by heat or chemicals. Now cheap, plastic, throw-away syringes can be used. They are pre-packed and then irradiated by an intense gamma ray source. This kills any bacteria but does not make the syringe radioactive.

Detecting radiation

Photographic fogging

Photographic film has a thin layer of silver-based chemical on the surface of the plastic or paper. Normally, this silver salt is affected by light falling on it – wherever it lands, it changes the chemical and blackens or fogs the film surface.

Alpha, beta or gamma radiation has a similar effect on this photographic emulsion, and so photographic film can be used to detect them. In fact, radioactive substances were first discovered, by accident, when Henri Becquerel, a French physicist, left some uranium rocks near photographic paper. He discovered that the paper had been blackened, This work led Pierre and Marie Curie to investigate these effects and gave us the term radioactivity.

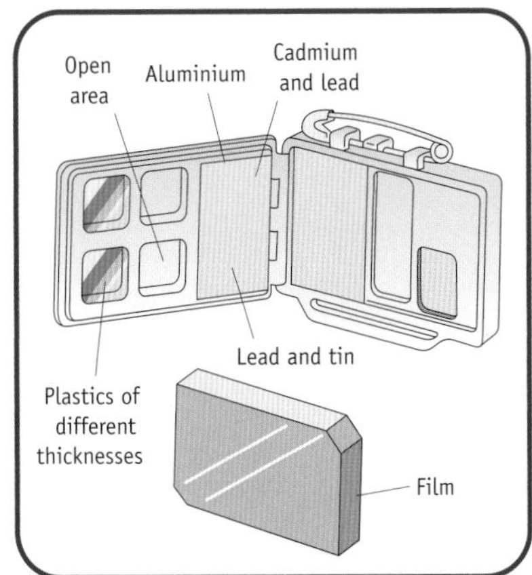

Figure 12.3 *A film badge, for indicating exposure of a worker to radioactivity*

Workers who use radioactive materials, particularly health workers in hospitals, must wear film badges throughout their working day. This enables a check to be made on the amount of radiation to which they have been exposed. When the film is developed, the amount of fogging gives a measure of the radiation exposure (Figure 12.3). Different windows are used to measure the amounts of the different types of radiation.

- A plastic window will absorb different energies of beta rays.
- Metal windows absorb different energies of gamma and X-rays.
- Aluminium will absorb low-energy X-rays.
- The other metals will absorb the high-energy X-rays.

Radiation damages you by ionising the cells in your body. If there is a lot of ionisation then there may be a path of ionisation through the cell wall. The cell wall may rupture and the cell can change into a cancerous one. This is a chance effect and the effect of the ionisation may cause no cancer.

Did you know?

Smoke Detectors

Most smoke detectors work on ionisation. There is a small amount, about 1/5000 of a gram, of Americium -241. This is a source of alpha particles. The ionisation chamber consists of two plates separated by about a cm with a battery across the plates. Alpha particles emitted from the source ionise the air particles in the chamber. The electrons are attracted to the positive terminal and the positive ions to the negative terminal. This causes a current between the plates. If smoke enters the ionisation chamber they attach themselves to the ions and neutralise them. The particles cannot reach the plate and the current drops. This drop in current triggers the alarm (Figure 12.4).

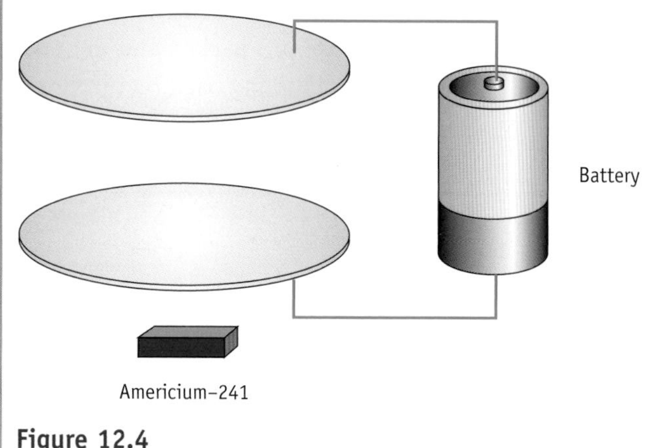

Figure 12.4

Scintillations

Some substances such as zinc sulphide are fluorescent. This means that they absorb radiation and give out energy again as a tiny burst of light. These flashes of light are called scintillations, and they may be observed by the naked eye or counted by a light detector and an electronic circuit. These scintillation counters are used in many modern instruments, including the gamma camera.

Using radiation in medicine

Treating cancer: radiation therapy

Radiotherapy is the treatment of cancers by radiation. Cancers are growths of cells which are out of control. Cancerous tumours can be treated by chemotherapy (very powerful drugs with side effects), surgery or radiation. The choice of treatment depends on the size and position of the tumour. Very often radiotherapy is used after surgery to destroy any remaining cancerous cells. The object of the radiation treatment is to cause damage to the cancer cells, which then stop reproducing. The tumour then shrinks.

Unfortunately, healthy cells can also be damaged by radiation. The amount of radiation has to be very accurately calculated so that sufficient damage is done to cancer cells without overdoing the damage to other cells. The radiation must be aimed very accurately at the tumour. This can be done using a simulator. A series of X-ray photographs are taken at different angles and a computer can build up a picture of the tumour and measure the amount of radiation to be given. Some localised tumours (e.g. a bone tumour) can be treated by irradiation with high-energy X-rays or gamma rays (Figure 12.5).

The gamma rays are emitted from a cobalt-60 source – a radioactive form of cobalt. The cobalt source is kept within a thick, heavy metal container. This has a slit in it to allow a narrow beam of gamma rays to emerge.

With this technique, the apparatus is arranged so that it can rotate around the couch on which the patient lies. This allows the patient to receive radiation from different directions. Rotating the source of radiation means that the tumour is **always** receiving radiation but the healthy tissue only receives a fraction of this radiation. Treatments are given as a series of small doses because tumour cells are killed more easily when they are dividing, and not all cells divide at the same time. This reduces the side effects, such as sickness.

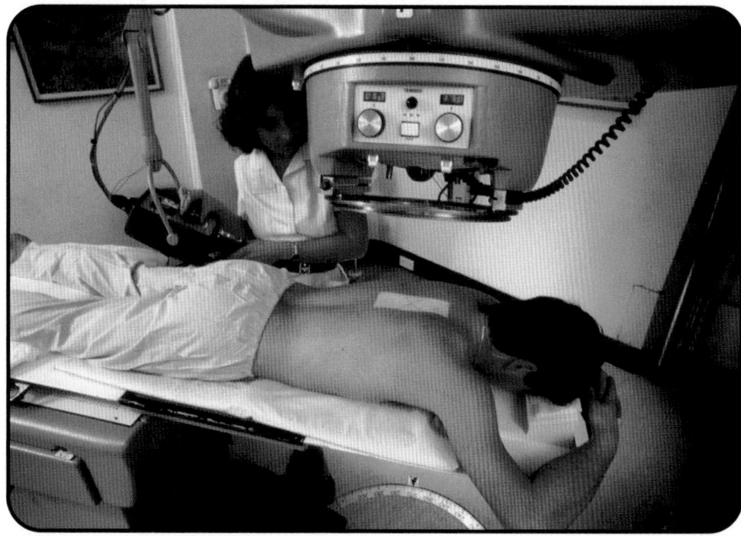

Figure 12.5 *Radiotherapy*

The gamma camera

It is important for doctors and scientists to be able to study internal organs without surgery. A **radiopharmaceutical** is used which can act as a tracer. The advantage of the radiopharmaceutical is that the radiation can be detected outside the body.

The radiopharmaceutical has two parts (Figure 12.6).

- A drug which is chosen for the particular organ that is being studied. Different organs have different drugs.
- A radioactive substance which is a gamma emitter.

Gamma is chosen since alpha or beta would be absorbed by tissue and would not be detected outside the body.

The combined substance is then injected into the patient.

To detect the radiation a **gamma camera** is used. This uses special crystals which emit flashes of light when the radiation reaches them. Photomultiplier tubes can change the light energy into electrical energy. The signals can then be displayed on a screen (Figure 12.7).

There are two types of studies:

- A static study is where there is a time delay between injecting the radioactive material and the build-up of radiation in the organ. This occurs in bone, lung or brain scans.
- A dynamic study is where the amount of radioactivity in an organ is measured as time increases. This occurs in the examination of the operation of the kidneys. This is called a renogram. The latter technique examines the working of the kidneys. The radioactive material reaches the kidneys due to the drug given to the patient. The radioactive material is taken out of the bloodstream by the kidneys. Within a few minutes of the drug being injected the radiation is concentrated in the kidneys. After 10–15 minutes, almost all the radiation should be in the bladder. The gamma camera takes readings every few seconds for 20 minutes. The computer adds up the radioactivity in each kidney. This can be shown as a graph of activity against time (Figure 12.8). Both kidneys take up the radioactivity. The radioactivity does not decrease as rapidly from the left kidney as it does from the right. The left kidney is not working correctly.

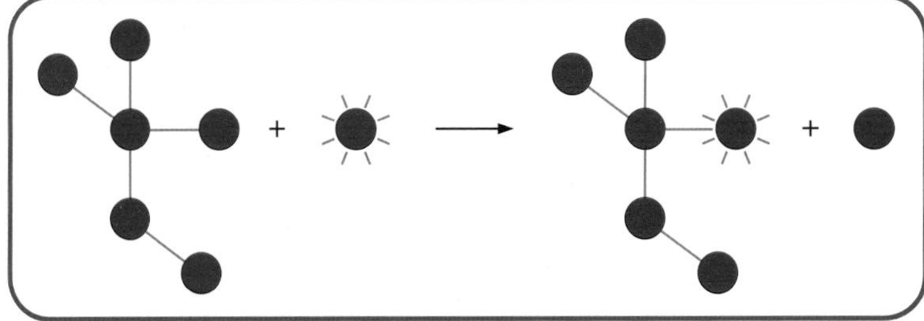

Figure 12.6 *Formation of a radiopharmaceutical*

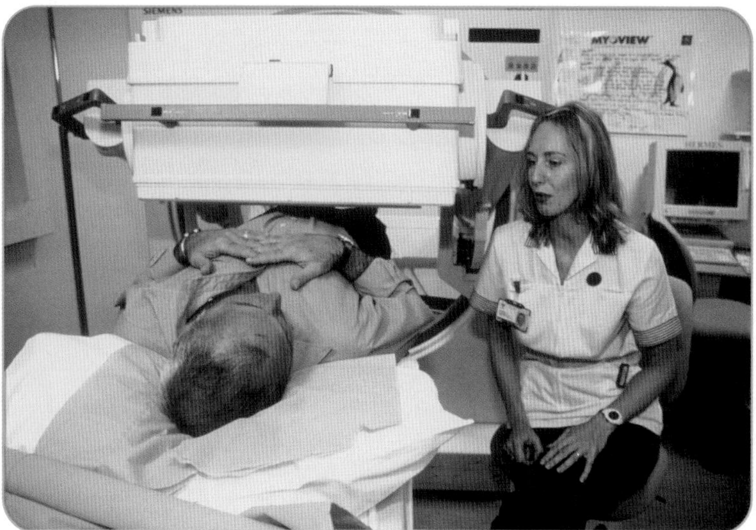

Figure 12.7 *A gamma camera*

Physics facts and key equations for ionising radiation

- The atom includes protons (+), neutrons and electrons (−).
- Radiation energy may be absorbed in a substance.
- Radiation can kill or change the nature of living cells.
- Alpha radiation is absorbed by paper and will travel about 5 cm in air. Beta is absorbed by aluminium and gamma radiation is reduced by lead or concrete.
- An alpha particle is a helium nucleus; a beta particle is a fast-moving electron and gamma radiation is a short-wave electromagnetic radiation.
- Ionisation is the gain or loss of an electron to produce a charged particle.
- Alpha particles produce much greater ionisation density than beta particles or gamma rays.
- Radiation is used in a detector of radiation.
- Radiation can destroy cells and is used in cancer treatment.
- Gamma cameras use radiation to examine the organs of the body.

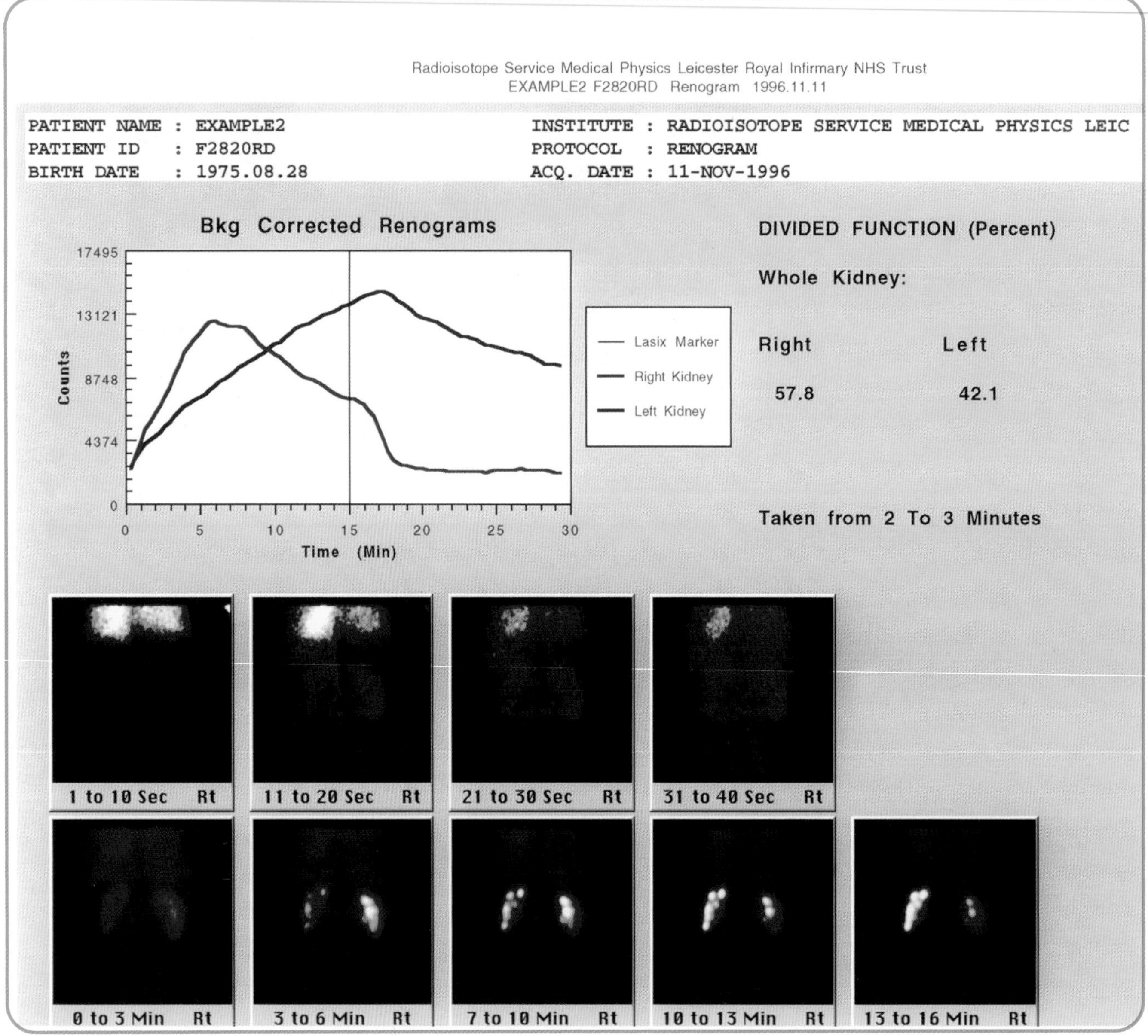

Figure 12.8 *A renogram. The renogram shows the time-activity for a kidney. It can be analysed to produce crucial information.*

Questions

1 a) State the three particles present in an atom and the charge on each particle.

b) Where is each of these particles located in the atom?

2 Alpha particles are emitted from a source.

a) (i) What is an alpha particle?

(ii) What is the range of alpha particles in air?

b) Ionisation occurs with alpha particles. What is meant by ionisation?

c) Why is it more dangerous to have an alpha source inside your body than a beta or gamma source?

3 Gamma radiation is used in medicine for many purposes.
   a) Describe how gamma radiation is used to detect if a patient's organ is working properly.
   b) Why is gamma radiation used rather than beta or alpha radiation?

4 a) What is a beta particle?
   b) Which material is usually used to absorb beta radiation?

5 Radiation can kill or cure in certain cases. Describe a use of radiation to cure a patient.

6 A patient is to be treated by radiation therapy.
   a) Which type of radiation is to be used if the radiation machine is outside the patient's body?
   b) Some tumours in the body are treated using a source of radiation that emits alpha radiation. The alpha emitters are placed inside the body.
      (i) Where should the radiation be placed if it is to destroy the tumour?
      (ii) Explain your answer.

13 Dosimetry

At the end of this chapter you should be able to...

1 State that the activity of a radioactive source is the number of decays per second and is measured in becquerels (Bq), where one becquerel is one decay per second.
2 Carry out calculations involving the relationship between activity, number of decays and time.
3 State that the absorbed dose *D* is the energy absorbed per unit mass of the absorbing material.
4 State that the gray is the unit of absorbed dose and that one gray is one joule per kilogram.
5 State that a radiation weighting factor is given to each kind of radiation as a measure of its biological effect.
6 State that the equivalent dose is the product of absorbed dose and radiation weighting factor and is measured in sieverts, Sv.
7 Carry out calculations involving the relationship between equivalent dose, absorbed dose and radiation weighting factor
8 State that the risk of biological harm from an exposure to radiation depends on
   (a) the absorbed dose
   (b) the kind of radiation
   (c) the body organs or tissue
9 Describe factors affecting the background radiation level.

Activity

All ionising radiation, that is, alpha, beta, gamma and X-rays, can cause damage to the cells of the body. It should be stressed that there is no minimum amount of radiation which is safe. The aim in physics is to measure the radiation and to estimate the risk when we are exposed to radiation. Some of this radiation occurs naturally but we still need an estimate of risk.

The becquerel

This unit measures the number of atoms (nuclei) which disintegrate or break up each second. If 50 nuclei break up each second then the activity of the source is 50 becquerels, abbreviated as Bq.

If A is the activity and N is the number of nuclei that disintegrate in a time of t seconds then

$$A = \frac{N}{t}$$

In practice, particularly in medical treatments, this is too small and larger units such as kBq and more often MBq (mega) are used. However, one gram of plutonium used in nuclear reactions has an activity of 2000 MBq.

Example

In a radioactive source 24 000 nuclei decay in one minute. What is the activity of the source?

Solution

N = 24 000 t = 60 s

$$A = \frac{N}{t}$$

$$= \frac{24000}{60}$$

$$= 400 \text{ Bq}$$

Example

An alpha source has an activity of 500 Bq. Calculate the total number of disintegrations that take place in 2 minutes.

Solution

A = 500 Bq and t = 120 s

$N = At$

$= 500 \times 120$

$= 60\ 000$

Example

A beta source is being used to treat a patient. It has an a activity of 6000 Bq. To treat the patient 5 400 000 nuclei in the beta source have to disintegrate. Calculate the time needed to treat the patient.

Solution

A = 6000 Bq N = 5 400 000

$$t = \frac{N}{A}$$

$$= \frac{5\,400\,000}{6000}$$

$$= 900 \text{ s}$$

Absorbed dose

When radiation reaches the body or tissue it is absorbed. This is called the absorbed dose (D) and is equal to one joule of energy absorbed by one kilogram of tissue.

This can be written as $D = \frac{E}{m}$

$$= \frac{\text{Energy absorbed by the material}}{\text{Mass of absorbing material}}$$

The unit is called the gray (Gy) and typical values are shown in the following table for a number of medical procedures.

Radiation treatment	Absorbed dose (Gy)
Chest X-ray	0.00015
CT scan	0.05
Gamma rays which would just produce reddening of the skin	3.0
Dose which if given to whole body in a short period would prove fatal in half the cases	5.0
Typical dose to a tumour over a six-week period	60.0

Example

A part of the body of mass 0.5 kg is exposed to radiation. The energy absorbed is 0.3 J.
Calculate the absorbed dose received by this part of the body.

Solution

$$D = \frac{E}{m}$$

E = 0.3 J   m = 0.5 kg

$$D = \frac{0.3}{0.5}$$

$$= 0.6 \text{ Gy}$$

Example

A patient receives a radiation scan for part of his body. He receives an absorbed dose of 0.05 Gy. The energy absorbed by the part of the body is 0.2 J. Calculate the mass of the body receiving the scan.

Solution

$$m = \frac{E}{D}$$

E = 0.2 J   D = 0.05 Gy

$$m = \frac{0.2}{0.05}$$

$$= 4 \text{ kg}$$

The absorbed dose has to be considered as a rough method of gauging response to radiation. For a typical person exposed to an absorbed dose of 10 Gy of radiation the effect could be fatal. Yet if received as heat energy your body temperature would only rise by a few thousandths of a degree celsius. At low absorbed doses of radiation such as those that we receive in everyday life, the cells in the body can repair themselves rapidly. At higher doses the cells may not be able to repair themselves and may be changed permanently or die. The permanently damaged cells may produce cancerous cells when they divide. Very few people have ever received a dose of greater than 2 Gy. If this occurred as a result of an accident then you would experience radiation sickness. The body's immune sytem would be damaged, and you might be vulnerable to other diseases.

In addition, radiation received by the body can produce a different effect depending on the type of radiation and the organ which receives the radiation.

Equivalent Dose

Tissue can receive the same amount of radiation but from different sources. In each case the effect on the tissue will be different. The risk of biological harm depends on

- the absorbed dose
- the kind of radiation
- the particular body organ or tissue exposed to the radiation.

To take account of the different types of radiation a number called the radiation **weighting factor** $\mathbf{w_R}$, is used.

The radiation weighting factors for X-rays, gamma and beta radiation is 1;

For fast neutrons it is 8

but for alpha radiation it is 20.

This is because the alpha radiation has a high ionisation density. When the radiation weighting factor is taken into account a quantity called the equivalent dose is measured. The unit of equivalent dose is sieverts (Sv) and is given the symbol *H*:

$$H = D\,\mathrm{w_R}$$

For each part of the body there will be different effects and the calculation of the absorbed equivalent dose takes into account the different types of tissue. For example, the lungs have a weighting of 0.2 and the bladder of 0.5. The use of these units can then allow us to estimate the risk of fatal cancer, which is defined as 5% per sievert. This means that if 100 people were exposed to an equivalent dose of 1 Sv to the whole body then five would develop a fatal cancer. In practice such exposures are not possible except in extreme accidents.

In research into radiation exposure, no definite health effects have been reported up to 100 mSv (0.1 Sv).

At 10 Sv, radiation sickness, skin burns and an increase in cancer can result.

Typical values of equivalent dose are shown below.

Investigation	Equivalent dose (mSv)
Chest X-ray	0.1
Spine X-ray	2.0
Stomach X-ray	4.0
CT scan	1 to 3.5
Bone scan	2.0
Annual exposure of aircraft crew	2.0
Renogram	2.0
Astronaut in space for one month	15

Example

A sample of tissue of mass 0.05 kg is is exposed to a source of radiation. The energy absorbed by the sample is 0.025 J. The weighting factor for the radiation is 20.

(a) Calculate the absorbed dose received by the sample of tissue.

(b) Calculate the equivalent dose received by the sample of tissue.

Solution

(a) $D = \frac{E}{m}$

$D = \frac{0.025}{0.05}$

$= 0.5$ Gy

(b) $H = D\,w_R$

$= 0.5 \times 20$

$= 10$ Sv

Example

A health worker is exposed to two types of radiation for the same length of time.

Information about the absorbed dose and the types of radiation is given in the table.

Type of Radiation	Absorbed dose (μ Gy)	Radiation weighting factor
Alpha	10	20
Gamma	25	1

Calculate the total equivalent dose for these radiations received by the health worker.

Solution

For alpha radiation

$H = D\,w_R$

$= 10 \times 10^{-6} \times 20$

$= 200 \times 10^{-6}$ Sv

For gamma radiation

$H = D\,w_R$

$= 25 \times 10^{-6} \times 1$

$= 25 \times 10^{-6}$ Sv

Total equivalent dose $H = 200 \times 10^{-6} + 25 \times 10^{-6}$ Sv

$= 225 \times 10^{-6}$ Sv

$= 225\ \mu$ Sv

It is important to put these figures in context. One mSv is about 100 times the radiation you experience when you travel by aircraft on holiday. But if you are part of the aircrew, then you will experience larger amounts due to the amount of travel. There are regulations about total flying times which take into account exposure to radiation, which will apply to all aircrew. It is also about half the annual equivalent dose of radiation which you will receive anyway from natural radiation.

Background radiation

There is radiation all around us. This is known as background radiation, and it is almost all from natural sources.

The table shows the typical equivalent dose we get every year from background radiation and manufactured sources of radiation. They are average figures, and vary a lot depending on our job and where we live.

Source	Annual dose (μSv)
Natural sources	
Radon and thoron gas from rocks and soil	800
Gamma rays from ground	400
Carbon and potassium in body	370
Cosmic rays at ground level	300
Total	**1870 μSv, that is, 1.87 mSv.**
Human made sources	
Medical uses: X-rays etc.	250
Fallout from weapons testing	10
Job (average)	5
Nuclear industry (e.g. waste)	2
Others (TV, aircraft flights , etc.)	11
Total	**278 μSv, that is, 0.278 mSv.**

Radon gas

This is radiation emitted from granite rock structures and is higher in some parts of the country than in others. The rocks contain radium which decays to produce radon gas. The design of modern houses actually prevents its dispersal easily since double glazing and the lack of chimneys means the gas cannot escape. The gas can be prevented from reaching the rooms by sealing the ground and making underfloor ventilation. In the USA the Environmental Protection Agency calls it the second largest cause of lung cancer.

Gamma rays

These are mainly emitted in substances like uranium and thorium which are contained in building materials.

Carbon and potassium

Some radioactivity occurs in air, food and water. The largest amount is from a radioactive source called potassium-40. The amount of radiation is controlled by our muscular systems. The average activity in our body is 4000 Bq.

Cosmic rays

This is radiation which comes from outer space. The Earth's atmosphere reduces our exposure to cosmic rays. However, these rays can penetrate buildings and enter our bodies.

Medical uses

The largest man-made radiation dose comes from medical sources. This covers X-rays, scans and cancer treatment. This figure is an average figure since unless you are receiving treatment you may receive only small amounts due to dental X-rays, for example.

Fallout

This is the name given to the radiation from nuclear weapons testing. Although there are very few tests taking place today, the radiation still exists from previous work.

Industry

Some jobs will involve the use of radiation. In heavy engineering, radiation is used to test welding joints. This technique is known as non-destructive testing. Miners also receive greater radiation due to thoron and radon gases under the ground. Aircrews, will also receive larger amounts since the thinner atmosphere will allow more radiation to reach them with greater exposure to cosmic rays.

Nuclear industry

Nuclear reactors will discharge small amounts of radiation into the atmosphere.

Others

Any TV monitor or computer monitor will emit very small amounts of radiation. All smoke detectors contain a radioactive source.

The different percentages that contribute to background radiation are shown in Figure 13.1. You can see that natural radiation is by far the biggest influence on us.

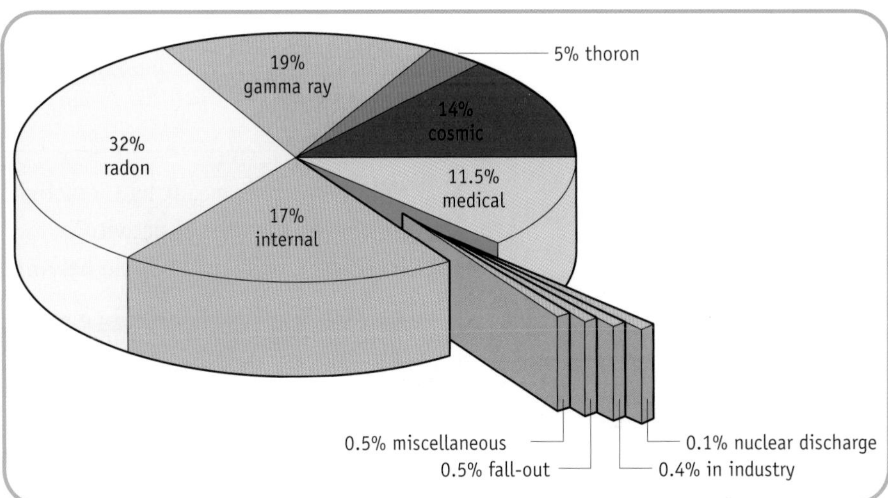

Figure 13.1 *Distribution of radiation*

The total annual equivalent dose in the UK averages about 2000 μSv, or about 2 mSv. But there is a big variation from person to person.

It is important to set the risk of death from radiation in context. A male aged 40 has a 1 in 500 risk of death from natural causes during a year. The table shows the annual risk of death from various causes. The risk from radiation is very small but it is not zero. Medical physicists aim to reduce the risk by using lower doses of radiation.

Activity	Average risk of death per year
Smoking 10 cigarettes per day	1 in 200
Being a male aged 40	1 in 500
Road accident	1 in 5000
Accident at work	1 in 20 000
Medical radiotherapy	1 in 250 000

Physics facts and key equations for dosimetry

- The activity of a radioactive source is measured in becquerels, where one becquerel is one decay per second.
- The absorbed dose D is the energy absorbed per unit mass of the absorbing material and is measured in grays, which are joules per kilogram.
- The risk of biological harm from radiation depends on:
  a) the absorbed dose
  b) the kind of the radiation
  c) the body organs or tissue exposed to the radiation.
- A radiation weighting factor w_R is given to each kind of radiation as a measure of its biological effect.
- Equivalent dose $H = D\, w_R$ and is measured in sieverts (Sv).
- There are several sources of background radiation, which can be split into natural and man-made.

Questions

1 a) State what is meant by the activity of a radioactive source.
b) What is the unit of activity?

2 Copy and complete the table below.

Number of disintegrations	Time (s)	Activity (Bq)
1.2×10^5	300	
800000	40	
2.5×10^4		5×10^6
24000		800
	600	1×10^6
	900	5×10^3

3 Copy and complete the table below.

Energy absorbed by material (J)	Mass of material (kg)	Absorbed dose (Gy)
0.05	0.02	
0.02	0.005	
	10	50×10^{-6}
	70	4.5×10^{-3}
0.8×10^{-6}		4×10^{-6}
0.1		2×10^{-3}

4 State the three factors which affect the risk of biological harm from radiation.
5 State the equation for equivalent dose and the unit in which it is measured.
6 A radioactive source emits radiation which is absorbed by a piece of tissue of mass 50 kg. The absorbed dose received is 6×10^{-4} Gy. The radiation weighting factor for this radiation is 10. Calculate the equivalent dose received by the tissue.
7 Copy and complete the table.

Absorbed dose (Gy)	Radiation weighting factor	Equivalent dose (Sv)
0.001	1	
0.003	20	
	20	0.002
	1	0.01
0.005		0.01
0.002		0.04

8 State two factors which contribute to background radiation. Explain whether these are natural or man made.

14 Half-life and safety

At the end of this chapter you should be able to...

1 State that the activity of a radioactive source decreases with time.
2 State the meaning of the term 'half-life'.
3 Describe the principles of a method for measuring the half-life of a radioactive source.
4 Carry out calculations to find the half-life of a radioactive isotope from appropriate data.
5 Describe the safety precautions necessary when handling radioactive substances.
6 State that the equivalent dose is reduced by shielding, by limiting the time of exposure or by increasing the distance from a source.
7 Identify the radioactive hazard sign and state where it should be displayed.

Half-life

When a radioactive substance disintegrates, the activity (number of disintegrations per second) depends only on the number of radioactive nuclei present, i.e. double the number, double the activity. The activity of all radioactive materials will decrease with time.

The half-life of a radioactive substance is the time taken for half the radioactive nuclei to disintegrate i.e. the time taken for the activity to fall by one half.

Half-life is measured in units of time. This could be in seconds, minutes, days or years.

Typical half lives are

- Uranium-238: 4.5×10^9 years
- Radium-226: 1600 years
- Cobalt-60: 5.3 years
- Sodium-24: 15 hours
- Copper-66: 5.2 minutes.

Total count rate and background count rate

In any radioactive experiment a count rate will be obtained even if no radioactive source is present. This is due to background radiation. The background count rate could be found by measuring the number of background counts in a known time and working out an average count rate (counts per second or counts per minute) from this.

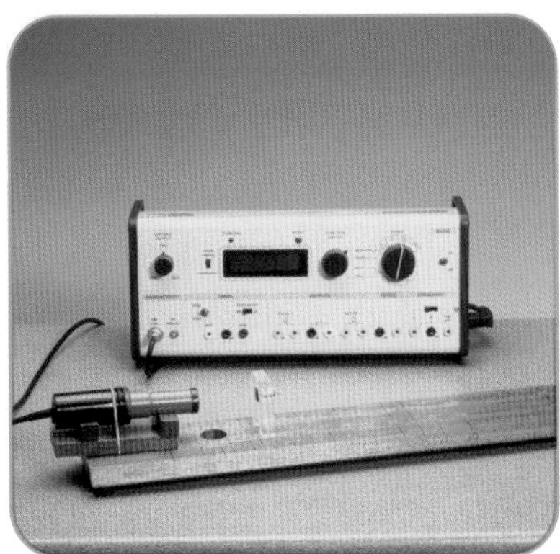

Figure 14.1 *Geiger Counter*

Measuring half-life

To measure the half-life of a radioactive source in a laboratory you will need the source and a detector/counter. This is shown in Figure 14.1. The detector is usually a Geiger counter.

Firstly you would take the background count rate several times without the source being present.

You would then calculate the average background count rate.

The radioactive source is now placed near the detector at a fixed distance and the count rate taken. Generally the counter will record for say 10 seconds, hold the reading for 2 seconds and then repeat this process. If you allow this to continue you will obtain a set of readings of total count rate at different times. Before you analyse the readings you will need to subtract the background count rate to obtain the true count rate due to the source alone.

Source count rate = total count rate – background count rate

In this experiment it is assumed that the source count rate represents the activity of the source. It may be that this experiment is demonstrated on a computer or shown on television since radioactive sources are not often stored in schools or colleges. You should then either calculate the half-life or draw a graph of source count rate against time, which will allow you to see the pattern more easily. In carrying out this experiment you should comply with the safety rules concerning radioactive sources, which are detailed later.

After one half-life ($t_{1/2}$), the source count rate drops to half the initial value. After a second half-life, the source count rate halves again – it is now one quarter of its original value. Three half-lives will see the source count rate reduce to one eighth, and so on.

A graph of source count rate against time is shown in Figure 14.2.

Average half life is 10 minutes.

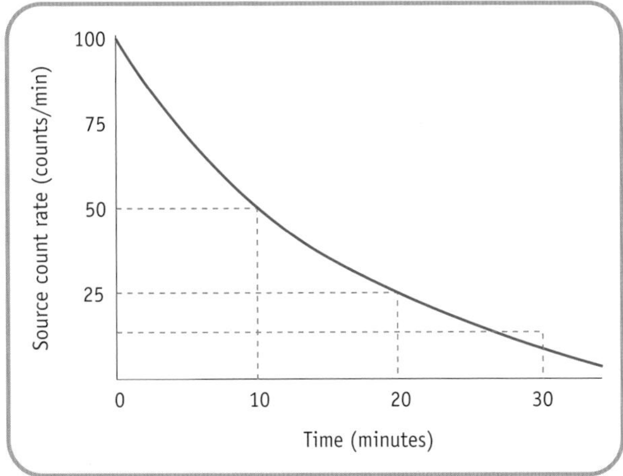

Figure 14.2 *A graph of count rate from a radioactive source against time*

Source count rate (counts per second)	Time (minutes)
100	0
50	10
25	20
12	30
7	40

Example

The activity of a radioactive source is 3200 Bq. After 150 minutes the activity is found to be 100 Bq. Calculate the half-life of the source.

Solution

Activity (Bq)	Number of half lives
3200	0
1600	1
800	2
400	3
200	4
100	5

5 half-lives = 150 minutes.
1 half-life = 30 minutes.

Example

A radioactive source has a half-life of 4 days. At the start of an experiment the total count rate recorded is 990 counts per minute. The background count rate is 30 counts per minute Find the total recorded count rate after 20 days.

Note:

1 Total count rate = count rate of source + background count rate.

2 The background rate is constant and does not decrease with time.

Solution

Count rate of source = 990 – 30 = 960 counts per minute

Source count rate (counts per minute)	Time (days)
960	0
480	4
240	8
120	12
60	16
30	20

After 20 days source count rate is 20 counts per minute
Total recorded count rate is 30 + 30 = 60 counts per minute.

Radiocarbon dating

Carbon-14 is a radioactive isotope which decays into nitrogen. It has a half-life of about 5500 years. All living things including people have a lot of carbon in them, a small amount of which will be carbon-14. When a plant or tree dies, the radioactive carbon-14 will decay and the amount of carbon-14 will decrease since in 5500 years there will be half as many atoms. The amount of carbon-14 in a sample can be measured and using some calculations the time of production of the sample can be found. Examples are of paper made from plants such as the Dead Sea scrolls or cloth such as the Turin Shroud thought to be the burial cloth of Christ but, using carbon dating, it has been shown that the Turin shroud was made about 700 years ago (Figure 14.3).

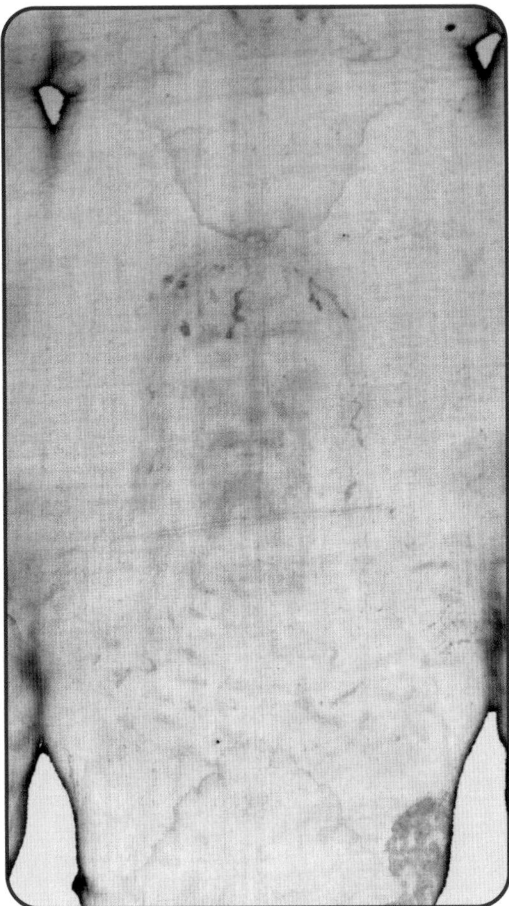

Figure 14.3 *Turin shroud*

Protection using radiation

Just after the discovery of radioactivity in 1896, radiation was thought to be good for health. Radium was advertised as a fertiliser for crops. Radium, now known as a major hazard, was sold in inhalers and an article in 1925 claimed that it had cured 27 illnesses from anaemia to sciatica. In addition during the 1920s the ladies who painted the luminescent faces on watches later developed cancer of the jawbone due to the practice of sucking their brushes. This led to changes in attitudes about radiation and steps were taken to reduce people's exposure to radiation.

There are still people who, as a result of their work, are exposed to ionising radiations. These people include medical workers using X-rays in hospitals, dentists and vets, research workers using radiation sources in their experiments and people who work in the nuclear power industry.

There are three methods by which radiation exposure can be reduced:

1 *Shielding a source of radiation with an appropriate thickness of absorber.* A radiographer wears a lead-lined apron (Figure 14.4). Radioactive sources are contained in lead containers.
2 *Limiting the time.* Sources should be moved and used as quickly as possible to reduce the radiation present. In the early days of radiation some workers changed photographic plates for X-rays with the beam still on. This resulted in radiation burns to the skin of the hand.
3 *Distance from source.* The further you are from the source the less radiation you will receive. In fact if you double the distance you will only receive a quarter of the radiation.

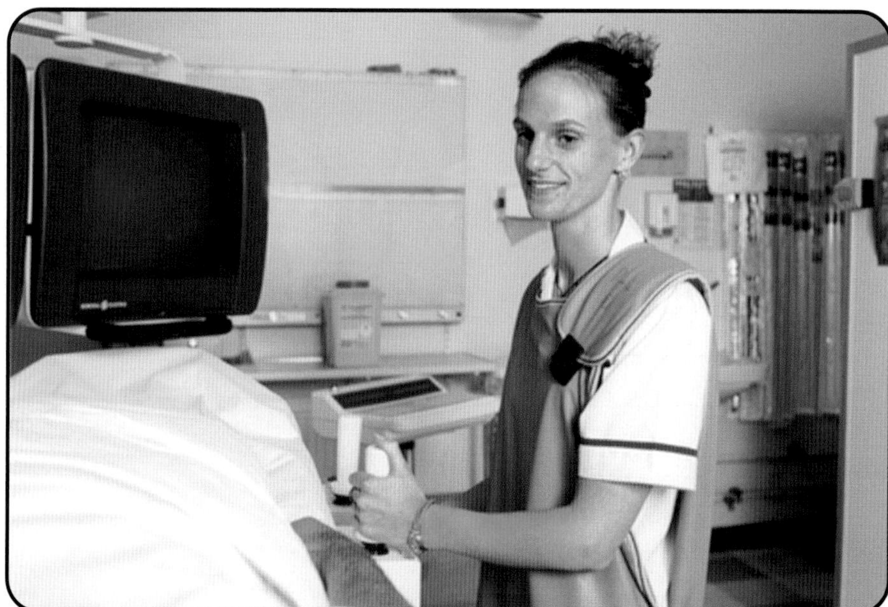

Figure 14.4 *A radiographer wearing a protective apron*

Safety with Radioactive Sources

1 Always use forceps or a lifting tool to remove a source (Figure 14.5). Never use bare hands.
2 Arrange a source so that its radiation window points away from the body.
3 Never bring a source close to your eyes for examination.
4 When in use, a source must always be attended by an authorised person and it must be returned to a locked and labelled store in its special shielded box immediately after use.
5 After any experiment with radioactive materials, wash your hands thoroughly before you eat.
6 In the UK, students under 16 years old may not normally handle radioactive sources.

The symbol for radiation sources being stored is shown in Figure 14.6. It must be displayed where radioactive materials are being used or stored, perhaps in a school or college. You will see it on doors in hospital areas and on boxes of radioactive sources which are being transported. It is an international symbol.

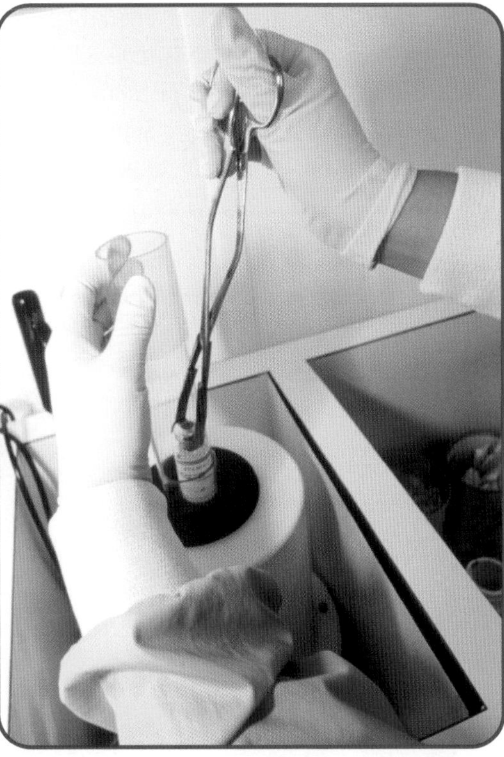

Figure 14.5 *Lifting a radioactive source with forceps*

Figure 14.6 *Radiation hazard symbol*

Did you know?

Generating radioactivity

The radioactive material used to carry out most nuclear medicine tests is Technetium-99, which is synthetic and does not occur naturally. While its 6-hour half-life is an advantage in reducing the radiation dose to the patient, it is not long enough to transport it from the manufacturer to the hospital. The answer is a generator. At a nuclear reactor, radioactive Molybdenum-99 is separated from the other materials in the used fuel and loaded on to a glass column packed with powder. The Molybdenum-99 has a 67 hour half-life, which is long enough for it to be delivered to the hospital and then decays to Technetium 99m – the radionuclide needed for the patient tests. On reaching the hospital salt solution is passed through the column. Molybdenum does not dissolve in salt but Technetium does. This is then used to make up a suitable drug (Figure 14.7).

Physics facts and key equations for half-life and safety

- The activity of a radioactive source decreases with time.
- The half-life of a radioactive substance is the time taken for half the radioactive nuclei to disintegrate.
- The safety precautions necessary when handling radioactive substances include handling the source with tongs, pointing the source away from you and not eating in the laboratory.
- The equivalent dose is reduced by shielding, by limiting the time of exposure or by increasing the distance from a source.
- The radioactive hazard sign is an international one and it should be displayed where sources are stored or are in daily use.

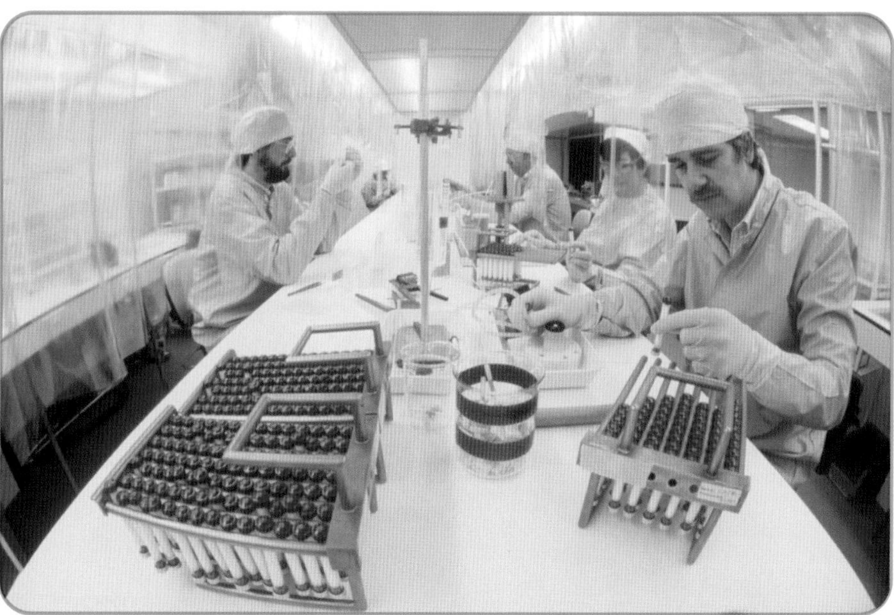

Figure 14.7 *Preparation of Technetium generator equipment*

Questions

1 What is meant by the statement that 'the half-life of a radioactive source is 6 hours'?

2 A radioactive source has an activity of 640 kBq. The half-life of the source is 8 days. What will be the activity of the source after 40 days?

3 A radioactive source has an activity of 600 kBq. In 24 hours the activity of the source is 75 kBq. Calculate the half-life of this source.

4 A source used in medicine has a half-life of 6 hours. The initial activity is 50 kBq. What is the source's activity after 36 hours?

5 Radioactive Iodine has a half-life of 8 days. After 24 days the sample of iodine has an activity of 1 k Bq. What was its initial activity?

6 Describe how you would measure the half-life of a radioactive source in a laboratory.
In your description state:
(i) the apparatus you would use
(ii) the measurements you would take
(iii) how you would use the measurements to calculate the half-life.

7 State two ways in which you could reduce the equivalent dose of radiation received by you.

8 State two places where you would see the sign displayed indicating radioactive hazards.

15 Nuclear reactors

At the end of this chapter you should be able to...

1. State the advantages and disadvantages of using nuclear power for the generation of electricity.
2. Describe in simple terms the process of fission.
3. Explain in simple terms a chain reaction.
4. Describe the principles of the operation of a nuclear reactor in terms of fuel rods, moderator, control rods, coolant and containment vessel.
5. Describe the problems associated with the disposal and storage of nuclear waste.

Production of electricity

Electricity is the main source of energy in all countries, both for domestic and business use. While coal, oil, solar, gas, hydro and other sources of energy can be used for this purpose, they all must be able to generate electricity to be of use to society. The main reasons for this is that electrical energy can be changed easily into other forms of energy, particularly heat and movement, and electrical energy can be sent over large distances by the national grid system.This is studied in the chapter on electricity, page 28.

However, the electricity must be produced at a power station, by one of a variety of methods. Renewable sources such as water, for hydro power, solar, wind and biomass such as wood are able to generate electrical power. The amounts generated will be small in relation to overall demand but they are useful as additional supplies at peak periods. They have the advantage that these sources can be used almost forever without any problems.

For large-scale generation of power, power stations can use a variety of sources. These will all have to create heat. In the case of coal, oil and gas they are burnt to create the heat to turn water into steam. Nuclear power operates in a different way to produce the heat energy.

Advantages of nuclear energy

- No gases are produced which cause problems with the environment such as carbon dioxide which still causes global warming.
- While uranium is a limited resource there are sufficient quantities in the world to ensure large amounts of fuel are available
- Unlike oil, nuclear power can produce the base load of constant power.

- Renewable sources such as wind, solar and biomass are useful, however, they cannot provide large amounts of reliable power at present.
- Oil may last for about a further 50 years and gas about 150, with coal possibly available for 300 years, but these fuels will become expensive to obtain. By comparison, a small amount of uranium, that is, a few kilograms, can supply several power stations.
- In the UK about 20% of energy is created from nuclear sources.

Disadvantages of nuclear energy

Figure 15.1 *Radioactive waste*

- The largest problem is the disposal of nuclear waste. Every power station will produce radioactive waste from the fuel rods. This waste has to be stored safely for hundreds of years. The amount of such waste is increasing (Figure 15.1).
- There is always a danger of accidents. The official statistics show that nuclear energy is the safest form of energy production, but accidents can happen. The worst accident was at Chernobyl in Russia in 1986 (Figure 15.2), when the workers ran the plant with the safety measures disconnected. The radiation from a damaged reactor was spread across a wide area of Europe by the wind and rain. Thirty-one people died from radiation-induced illnesses and 135,000 people living near the reactor had to leave the area. Overall the accident released radiation with a total activity of 1.85×10^{18} Bq. 134 people showed signs of acute radiation sickness immediately after the accident. The effects on the population are still continuing since firefighters and other emergency service people were exposed to radiation.

Figure 15.2 *Chernobyl after the accident in 1986*

Fission

Nuclear power stations use uranium-235 as a fuel. To produce the energy required, the nuclei of uranium are bombarded by neutrons. If the uranium nucleus absorbs a neutron, the nucleus can become unstable. The nucleus splits into two smaller pieces, as shown in Figure 15.3. This process is called **fission**. The process releases energy and further neutrons.

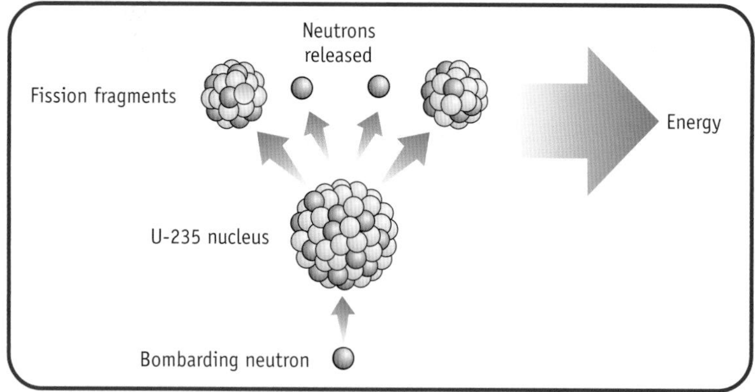

Figure 15.3 *Fission*

These neutrons can go on to hit further uranium nuclei, which split and again release energy and more neutrons. This process can continue and is called a **chain reaction** (Figure 15.4). Such a reaction will take place in an atomic bomb and the amount of material needed to produce this is called the critical mass. In a nuclear plant the reaction must be controlled so that the amount of power produced is sufficient for our needs.

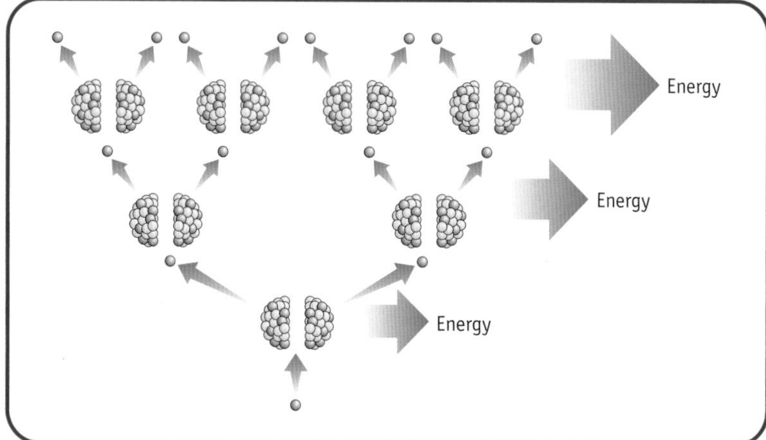

Figure 15.4 *Chain reaction*

A nuclear power station

A nuclear reactor is illustrated Figure 15.5. The key elements are as follows.

The fuel rods

These contain natural uranium which is enriched so that fission can occur. The uranium is found as uranium oxide and after being purified it is called 'yellowcake'. Uranium-238 is the most common form of uranium but uranium-235 is the form needed to produce fission. The amount of uranium in a fuel rod is well below the critical mass so that an explosion cannot naturally occur. The fuel rods will have to be replaced every few years. This is done by a machine.

Graphite moderator

When the neutrons are emitted after fission they are moving very fast. They will not be able to be 'captured' by other nuclei so fission will not occur. However, if they are slowed down then there is a greater chance that fission will occur. This is done using a moderator. This is generally graphite. When the neutrons collide with the graphite atoms they slow down and can now cause fission to take place in other uranium nuclei. Neutrons will leave one rod and cause nuclei to split in another nearby rod.

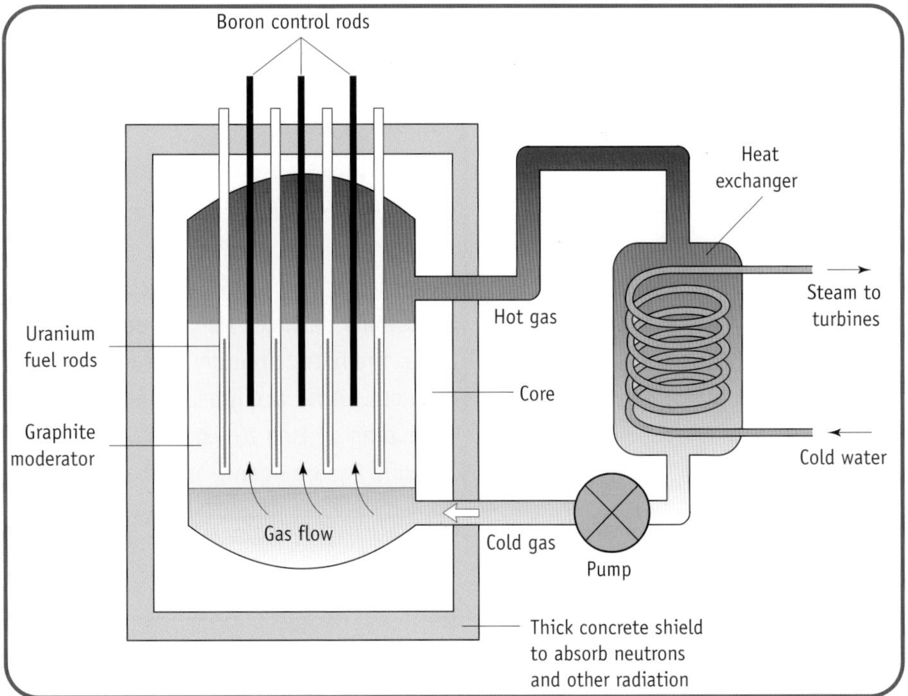

Figure 15.5 *A nuclear reactor*

Control rods

The amount of electrical power required will vary, with peak demands during the day, whereas during the night the demand is lower.

- Rods of boron can absorb additional neutrons and control the number available for fission.
- They can be raised and lowered into the reactor as necessary.
- They also provide a safety feature since in the event of an accident all the control rods would be lowered automatically to absorb the neutrons.

Coolant

The heat produced during this process must be removed from the reactor. This is normally done in UK reactors using carbon dioxide gas. This circulates around the core of the reactor and is continually heated and then passes the heat to water through a heat exchanger. The water is not radioactive and is able to be reused. The water is then changed to steam, which is used to drive a turbine, which turns on the generators to produce electricity. This last stage is similar in any kind of power station after the heat is produced.

Containment vessel

The key parts of a nuclear reactor, which form the core, are contained in a containment vessel. This is designed so that no radiation can escape. It is generally several metres thick and has a concrete top. It is safe to walk on the top of the reactor but access is almost impossible unless it has to be repaired. The vessel can withstand a small conventional explosion.

Did you know?

The boy who built his own reactor

A 17 year old American boy called David Hahn built his own reactor in 1995, in a shed in his garden. The reactor was made with Radium from old paint and Americium from smoke detectors. These provided a source of alpha particles. The alpha particles struck Beryllium to produce fast neutrons. The neutrons then irradiated Thorium and Uranium, which were obtained from various sources. David acquired the information to build it from many different places including writing to various organisations. The 'reactor' was disposed of by the Atomic Energy Agency in America.

Disposal of nuclear waste

Nuclear waste is split into different categories.

High-level waste is mainly spent nuclear fuel. After several years of use the fuel rods are taken out and sent to a reprocessing plant. These factories extract the useful parts of the fuel, which can be used to make new fuel rods. Unfortunately the material that remains is very highly radioactive. The materials are stored initially in a large tank of water. This absorbs the heat and also some of the radiation. After about a year the materials can be handled but are still very radioactive. The materials have very long half-lives and have to be stored in a suitable environment which is safe to human beings for a long time. Initially they were stored in large water-cooled tanks but these have further problems with leaks and the cooling process. The latest suggestion is to change the materials into a form of glass and then store it in boreholes deep in the ground. This has further problems since no one knows if the glass-like materials could crack, or if the radiation could leak into rocks and possibly into water supplies. This is still not considered the full solution and the waste is at present stored on site at the reprocessing plants. Part of the problem is the length of time before the waste materials can be considered safe since scientists cannot guarantee storage of anything over a long period (Figure 15.6).

Figure 15.6 *Nuclear low-level waste storage facility*

Other suggestions have been to put the waste in a spacecraft and send it towards the sun – but what if the spacecraft explodes on launch?

Low-level waste is the remainder of nuclear wastes and those generated by hospitals etc. 'Low-level' does not mean it is not dangerous. It means that the radioactivity is less concentrated but the waste still has to be stored. It used to be dumped at sea but this is now banned since humans cannot continually pollute other parts of the environment. In a typical country, since nuclear reactors began in 1957 the nuclear waste amounts to about 9000 000 kg but the total amount of poisonous waste is about 50 thousand million kg

In summary, the storage of the waste centres on two issues:

- The disposal technique. An example might be the waste is made into a form of glass or stored in large tanks.

- The possible sites for the storage. These tend to be concentrated in certain parts of the country but with the amount of waste increasing this is a problem for current and future generations.

There is a tendency to think that nuclear waste is not our problem but someone else's problem. Since we all use electricity, some of which is nuclear generated, it is all our problem.

Did you know?

Transporting nuclear waste

Nuclear fuel is transported around the country, mainly by rail. To demonstrate the safety of the containers a rail crash was organised. A train with three carriages was travelling at 100 mph. This train struck a large nuclear flask as it lay on the track. The train and carriages were badly damaged but the flask remained intact. There have been no reported leaks from any of these flasks (Figure 15.7).

Figure 15.7 *Safety demonstration of the transportation of nuclear fuel by train*

Physics facts and key equations for nuclear reactors

- There are advantages and disadvantages of using nuclear power for the generation of electricity.
- Fission occurs when certain nuclei are bombarded by neutrons and become unstable. The nuclei break into smaller pieces and release energy.
- A chain reaction occurs when neutrons released by fission go on to strike more nuclei and cause more nuclei to split.
- A nuclear reactor has fuel rods, a moderator, control rods, coolant and a containment vessel.
- The disposal and storage of nuclear waste are difficult due to the long half-life of some sources and the long-term storage that is required.

Questions

1 State two advantages and two disadvantages in using nuclear power to generate electricity.

2 Fission occurs when a neutron hits certain nuclei. Describe what happens next.

3 What is the function of the control rods and why are they made of a substance such as Boron?

4 A nuclear reactor has several parts. Explain the function of each part listed below;
fuel rods moderator coolant concrete shield

5 Several sites have been proposed which can be used to store nuclear waste.
a) State two reasons why the disposal of nuclear waste at sites can cause problems.
b) When sites are chosen state two difficulties in storing the waste.

6 The data below is about a nuclear reactor
Number of fuel rods in reactor = 1700
Mass of one fuel rod = 14 kg
Amount of uranium-235 in one rod = 3 %
Energy produced by 1 kg of uranium-235 = 1×10^{14} J
a) Calculate the mass of uranium-235 in the fuel rods.
b) If there is a fission of all the uranium-235 nuclei, calculate the total amount of energy that is released.

Exam Questions

1 A radioactive source emits radiation with a half-life of one year.

a) Explain what is meant by the term half-life.

b) The source has an activity of 1 MBq. What will be its activity in 6 years?

c) When radiation is used for treating humans, the effect of the radiation will depend on three factors. State the three factors.

d) Radiation of radiation weighting factor 20 has an energy of 0.35 J. It is absorbed by a piece of tissue of mass 70 kg.
Calculate
(i) the absorbed dose received by the tissue
(ii) the equivalent dose received by the tissue.

2 Nuclear reactors are the main method by which electricity is produced in the UK.

a) Nuclear reactors operate using fission. Explain clearly what is meant by this process. Start your description with the sentence
'When a neutron is fired at an uranium nucleus'

b) Describe two advantages and two disadvantages of using nuclear energy.

c) The production of this energy depends on **nuclear fission** and a **controlled chain reaction**. Explain what is meant by these terms.

d) Radioactive waste is a problem associated with nuclear reactors. Describe two issues in the disposal of high level waste.

3 During radiation treatment, the amount of ionisation produced by different radiations is measured.

a) Explain what is meant by ionisation.

b) Which type of radiation produces the greatest amount of ionisation in a fixed volume.

c) During an examination of a patient a gamma source is used as a tracer. This source is mixed with a drug and put into the patient's bloodstream. The radiation is then monitored outside the body.
(i) Why is a gamma source used rather than alpha or beta sources?
(ii) The half-life of the gamma source is 6 hours. Why is this half-life used rather than one which has longer or short half-life?

Relationships required for Intermediate 2 Physics

$d = \bar{v}t$

$s = \bar{v}t$

$a = \dfrac{\Delta v}{t}$

$a = \dfrac{v - u}{t}$

$W = mg$

$F = ma$

$p = mv$

$E_w = Fd$

$E_p = mgh$

$E_k = \dfrac{1}{2}mv^2$

$P = \dfrac{E}{t}$

$percentage\ efficiency = \dfrac{useful\ E_o}{E_i} \times 100$

$percentage\ efficiency = \dfrac{useful\ P_o}{P_i} \times 100$

$E_h = cm\Delta T$

$E_h = ml$

$Q = It$

$V = IR$

$R_T = R_1 + R_2 + \ldots$

$\dfrac{1}{R_T} = \dfrac{1}{R_1} + \dfrac{1}{R_2} + \ldots$

$V_2 = \left(\dfrac{R_2}{R_1 + R_2}\right)V_S$

$\dfrac{V_1}{V_2} = \dfrac{R_1}{R_2}$

$P = IV$

$P = I^2R$

$P = \dfrac{V^2}{R}$

$\dfrac{n_s}{n_p} = \dfrac{V_s}{V_p} = \dfrac{I_p}{I_s}$

$V_{gain} = \dfrac{V_o}{V_i}$

$P_{gain} = \dfrac{P_o}{P_i}$

$v = f\lambda$

$P = \dfrac{1}{f}$

$A = \dfrac{N}{t}$

$D = \dfrac{E}{m}$

$H = Dw_R$

Symbols and Units for Terms in Intermediate 2 Physics

Term	Symbol	Unit	Unit Abbreviation
distance	*s* or *d*	metre	m
displacement	*s*	metre	m
speed, velocity	*v*	metre per second	m/s
time	*t*	second	s
change of velocity	Δv	metre per second	m/s
average velocity	$\bar{v}$	metre per second	m/s
initial velocity	*u*	mtre per second	m/s
final velocity	*v*	metre per second	m/s
acceleration	*a*	metre per second per second	m/s^2
mass	*m*	kilogram	kg
weight	*W*	newton	N
force	*F*	newton	N
acceleration due to gravity	*g*	metre per second per second	m/s^2
gravitational field strength	*g*	newton per kilogram	N/kg
momentum	*p*	kilogram metre per second	kg m/s
energy	*E*	joule	J
work done	*W* or E_w	joule	J
potential energy	E_p	joule	J
height	*h*	metre	m
kinetic energy	E_k	joule	J
power	*P*	watt	W
efficiency	(η)	-	-
temperature	*T*	degree Celsius	°C
specific heat capacity	*c*	joule per kilogram per degree Celsius	J/kg °C
specific latent heat	*l*	joules per kilogram	J/kg
heat energy	E_h	joule	J
electric charge	*Q*	coulomb	C
electric current	*I*	ampere	A
voltage, potential difference	*V*	volt	V
supply voltage	V_s	volt	V

Term	Symbol	Unit	Unit Abbreviation
resistance	R	ohm	Ω
total resistance	R_T	ohm	Ω
number of turns on primary coil	n_p	-	-
number of turns on secondary coil	n_s	-	-
primary voltage	V_p	volt	V
secondary voltage	V_s	volt	V
primary current	I_p	ampere	A
secondary current	I_s	ampere	A
input voltage	V_i	volt	V
output voltage	V_o	volt	V
voltage gain	A_o or V_{gain}	-	-
wavelength	λ	metre	m
frequency	f	hertz	Hz
period	T	second	s
amplitude	A	metre	m
angle	θ	degree	°
critical angle	θ_c	degree	°
power (of a lens)	P	dioptre	D
focal length	f	metre	m
activity	A	becquerel	Bq
count rate	-	counts per second (counts per minute)	-
absorbed dose	D	gray	Gy
radiation weighting factor	w_R	-	-
equivalent dose	H	sievert	Sv
half-life	$t_{1/2}$	second (minute, hour, day, year)	s

Data Sheet for Intermediate 2 Physics

Speed of light in materials

Material	Speed (m/s)
Air	$3{\cdot}0 \times 10^8$
Carbon dioxide	$3{\cdot}0 \times 10^8$
Diamond	$1{\cdot}2 \times 10^8$
Glass	$2{\cdot}0 \times 10^8$
Glycerol	$2{\cdot}1 \times 10^8$
Water	$2{\cdot}3 \times 10^8$

Gravitational field strengths

	Gravitational field strength on the surface (N/kg)
Earth	10
Jupiter	26
Mars	4
Mercury	4
Moon	1.6
Neptune	12
Saturn	11
Sun	270
Venus	9

Specific latent heat of fusion of materials

Material	Specific latent heat of fusion (J/kg)
Alcohol	$0{\cdot}99 \times 10^5$
Aluminium	$3{\cdot}95 \times 10^5$
Carbon dioxide	$1{\cdot}80 \times 10^5$
Copper	$2{\cdot}05 \times 10^5$
Iron	$2{\cdot}67 \times 10^5$
Lead	$0{\cdot}25 \times 10^5$
Water	$3{\cdot}34 \times 10^5$

Specific latent heat of vaporisation of materials

Material	Specific latent heat of vaporisation (J/kg)
Alcohol	$11{\cdot}2 \times 10^5$
Carbon dioxide	$3{\cdot}77 \times 10^5$
Glycerol	$8{\cdot}30 \times 10^5$
Turpentine	$2{\cdot}90 \times 10^5$
Water	$22{\cdot}6 \times 10^5$

Speed of sound in materials

Material	Speed (ms)
Aluminium	5200
Air	340
Bone	4100
Carbon dioxide	270
Glycerol	1900
Muscle	1600
Steel	5200
Tissue	1500
Water	1500

Specific heat capacity of materials

Materials	Specific heat capacity (J/kg°C)
Alcohol	2350
Aluminium	902
Copper	386
Glass	500
Ice	2100
Iron	480
Lead	128
Oil	2130
Water	4180

Melting and boiling points of materials

Material	Melting point (°C)	Boiling point (°C)
Alcohol	−98	65
Aluminium	660	2470
Copper	1077	2567
Glycerol	18	290
Lead	328	1737
Iron	1537	2747

Radiation weighting factors

Type of radiation	Radiation weighting factor
alpha	20
beta	1
fast neutrons	10
gamma	1
slow neutrons	3

Page numbers in italics indicate illustrations not included in text page range.

Answers to Unit 1

Chapter 1

1 $Q = It$

$Q = 10 \times (2 \times 60)$

$Q = 1200\ C$

2 $Q = It$

$500 = 2.0 \times t$

$t = \frac{500}{2.0} = 250\ s$

3 $Q = It$

$150 = I \times (5 \times 60)$

$I = \frac{500}{300} = 0.5\ A$

4 9 J of energy is given to each coulomb of charge.

5 See Figure 1.4 on page 5.

6 $A_1 = 0.5$ A since current is the same at all points in a series circuit.

$A_2 = 3.0$ A as circuit current = sum of currents in the branches of a parallel circuit.

$A_3 = 1.5$ A as circuit current = sum of currents in the branches of a parallel circuit.

$V_1 = 2.0$ V as supply voltage = sum of voltages in a series circuit.

$V_2 = 6$ V as voltages are the same across components connected in parallel.

$V_3 = 9$ V as supply voltage = sum of voltages in a series circuit.

7 $R_T = R_1 + R_2 + R_3$

$R_T = 47 + 100 + 150$

$R_T = 297\ \Omega$

8 $\frac{1}{R_T} = \frac{1}{R_1} + \frac{1}{R_2} + \frac{1}{R_3} = \frac{1}{20} + \frac{1}{20} + \frac{1}{10}$

$= 0.05 + 0.05 + 0.1 = 0.2$

$\frac{1}{R_T} = 0.2$

$R_T = \frac{1}{0.2} = 5\ \Omega$

9 For the two resistors connected in parallel:

$\frac{1}{R_p} = \frac{1}{R_1} + \frac{1}{R_2} = \frac{1}{60} + \frac{1}{30} = 0.017 + 0.033 = 0.05$

$\frac{1}{R_p} = 0.05$

$R_p = \frac{1}{0.05} = 20\ \Omega$

$R_{XY} = 30 + 20 + 20 = 70\ \Omega$

10 See Figures 1.5 and 1.6 on page 6.

11 $V = IR$

$V = 0.012 \times 1000$

$V = 12\ V$

12 $V = IR$

$230 = 4.5 \times R$

$R = \frac{230}{4.5} = 51\ \Omega$

13 $V = IR$

$4.5 = I \times 18$

$V = \frac{4.5}{18} = 0.25\ A$

14 $R_T = R_1 + R_2$

$R_T = 1000 + 3000 = 4000\ \Omega$

$I\text{circuit} = \frac{V_s}{T_T} = \frac{9}{4000} = 0.00225\ A$

$V_{3k\Omega} = I_{circuit} \times R_{3k\Omega} = 0.00225 \times 3000 = 6.75\ V$

Chapter 2

1 (a) $P = IV$

$P = 5.5 \times 230$

$P = 1265$ W

(b) $P = \frac{E}{t}$

$1265 = \frac{E}{1}$

$E = 1265$ J

2 (a) Electrical to heat and light

(b) $V = IR$

$12 = I \times 3$

$I = \frac{12}{3} = 4$ A

(c) $P = IV$

$P = 4 \times 12$

$P = 48$ W

3 $P = \frac{V^2}{R}$

$P = \frac{230^2}{17.6}$

$P = 3006$ W

4 $P = \frac{V^2}{R}$

$50 = \frac{12^2}{R}$

$R = \frac{144}{50} = 2.88\ \Omega$

5 $P = I^2R$

$138 = 0.6^2 \times R$

$R = \frac{138}{0.36} = 383\ \Omega$

6 Any value over 230 V e.g. 320V.

7 50 Hz

Chapter 3

1 Increase strength of magnetic field of magnet, increase number of loops of wire on coil and increase speed of movement of coil.

2 $\frac{N_s}{N_p} = \frac{V_s}{V_p}$

$\frac{N_s}{40} = \frac{120}{10}$

$N_s = 12 \times 40 = 480$ turns

3 $\frac{V_s}{V_p} = \frac{N_s}{N_p}$

$\frac{V_s}{120} = \frac{400}{8000}$

$V_s = 0.05 \times 120 = 6$ V

4 $\frac{V_p}{V_s} = \frac{N_p}{N_s}$

$\frac{V_p}{5} = \frac{1500}{750}$

$V_p = 2 \times 5 = 10$ V

5 $\frac{V_s}{V_p} = \frac{N_s}{N_p} = \frac{I_p}{I_s}$

$\frac{I_p}{I_s} = \frac{N_s}{N_p}$

$\frac{I_p}{1} = \frac{750}{250}$

$I_p = 3$ A

6 (a) a.c.

(b) $\frac{V_s}{V_p} = \frac{N_s}{N_p} = \frac{I_p}{I_s}$

$\frac{I_p}{I_s} = \frac{V_s}{V_p}$

$\frac{I_p}{2} = \frac{10}{230}$

$I_p = 0.043 \times 2 = 0.086$ A

(c) $\frac{N_s}{N_p} = \frac{V_s}{V_p}$

$\frac{N_s}{8280} = \frac{10}{230}$

$N_s = 0.043 \times 8280 = 360$ turns

Chapter 4

1 e.g. loudspeaker – electrical to sound, filament lamp – electrical to heat and light, and LED – electrical to light

2 (a) Figure 4.3 + switch connected in series

(b) Since components are connected in series then

Supply voltage = voltage across LED + voltage across resistor

$6 = 1.75 + V_R$

$V_R = 4.25$ V

$V_R = IR$

$4.25 = 11 \times 10^{-3} \times R$

$$R = \frac{4.25}{11 \times 10^{-3}} = 386\ \Omega$$

3 (a) sound to electrical energy (b) heat to electrical energy, (c) light to electrical energy

4 (a) $V = IR$

$6 = 12 \times 10^{-3} \times R$

$$R = \frac{6}{12 \times 10^{-3}} = 500\ \Omega$$

(b) A value greater than 12 mA e.g. 13 mA

5 As the temperature of the oven increases the resistance of the thermistor decreases and the voltage across the thermistor decreases. The voltage across the resistor R increases and when it is above a certain value the transistor switches on and the lamp lights.

6 (a) Figure 4.12 (b) Figure 4.11, both on page 40.

7 (a) $\text{Voltage gain} = \frac{\text{output voltage}}{\text{input voltage}}$

$$\text{Voltage gain} = \frac{3.0}{10 \times 10^{-3}} = 300$$

(b) same

8 (a) $\text{Voltage gain} = \frac{\text{output voltage}}{\text{input voltage}}$

$$200 = \frac{\text{output voltage}}{3 \times 10^{-3}}$$

Output voltage $= 200 \times 3 \times 10^{-3} = 0.6$ V

Unit 1 Exam Questions

1 (a) (i) $R_T = R_1 + R_2 = 36 + 18 = 54\ \Omega$

$V_S = I_{circuit} R_T$

$9 = I \times 54$

$$I_{circuit} = \frac{9}{54} = 0.167 \text{ A}$$

(a) (ii) $V_{18\Omega} = I_{circuit} R_{18\Omega} = 0.167 \times 18 = 3$ V

(b) (i) $$\frac{1}{R_{AB}} = \frac{1}{R_1} + \frac{1}{R_2} + \frac{1}{R_3} = \frac{1}{100} + \frac{1}{75} + \frac{1}{100}$$

$= 0.01 + 0.013 + 0.01 = 0.033$

$$R_{AB} = \frac{1}{0.033} = 30\ \Omega$$

(b) (ii) $$P = \frac{V^2}{R} = \frac{230^2}{30} = 1763 \text{ W}$$

2 (a) To ensure correct voltage is applied to each lamp.

(b) $P = IV$

$3 = I \times 12$

$$I = \frac{3}{12} = 0.25 \text{ A}$$

(c) Since lamps and resistor are in series then:

Supply voltage = 19 × voltage across each lamp + voltage across resistor

$230 = (19 \times 12) + V_R$

$230 = 228 + V_R$

$V_R = 2$ V but $V_R = IR$

$2 = 0.25 \times R$

$$R = \frac{2}{0.25} = 8\ \Omega$$

(d) $Q = It = 0.25 \times 60 = 15$ C

(e) $V = IR$

$12 = 0.25 \times R$

$$R = \frac{12}{0.25} = 48\ \Omega$$

3 (a) (i) $P = \frac{E}{T}$

$$2116 = \frac{E}{100}$$

$E = 2116 \times 180 = 3.81 \times 10^5$ J

(a) (ii) In the resistance wire or element

(b) $P = \frac{V^2}{R}$

$$2116 = \frac{230^2}{R}$$

$$R = \frac{52\,900}{2116} = 25\ \Omega$$

4 (a) LED

(b) To prevent damage to the LED from too high a current in the LED (or too high a voltage across the LED).

(c) Since components are connected in series then:
Supply voltage = voltage across LED + voltage across resistor

$10 = 1.8 + V_R$

$V_R = 8.2$ V

$V_R = IR$

$8.2 = 11 \times 10^{-3} \times R$

$$R = \frac{8.2}{11 \times 10^{-3}} = 745\ \Omega$$

5 (a) $\frac{V_p}{V_s} + \frac{N_p}{N_s}$

$$\frac{V_p}{12} = \frac{2000}{200}$$

$V_p = 10 \times 12 = 120$ V

(b) $P = IV$

$24 = I \times 12$

$$I = \frac{24}{12} = 2\ \text{A}$$

(c) $\frac{I_p}{I_s} + \frac{N_s}{N_p}$

$$\frac{I_p}{2} = \frac{200}{2000}$$

$I_p = 0.1 \times 2 = 0.2$ A

6 (a) Circuit A

(b) (i) $R_T = R_1 + R_2 = 100 + 100 = 200\ \Omega$

(ii) $\frac{1}{R_T} = \frac{1}{R_1} + \frac{1}{R_2} = \frac{1}{100} + \frac{1}{100}$

$= 0.01 + 0.01 = 0.02$

$$R_{AB} = \frac{1}{0.02} = 50\ \Omega$$

(c) (i) $V = IR$

$230 = I \times 200$

$$I = \frac{230}{200} = 1.15\ \text{A}$$

(ii) $V = IR$

$230 = I \times 50$

$$I = \frac{230}{50} = 4.6\ \text{A}$$

(d) Circuit B – it has the higher power rating since lower resistance means higher current reading for the same voltage.

7 (a) n-channel enhancement MOSFET (b) As the temperature of the thermistor increases, the resistance of the thermistor decreases and so the voltage across the thermistor decreases. The voltage across the variable resistor must therefore increase and the MOSFET switches on and the LED lights. (c) Change the positions of the thermistor and the variable resistor.

Answers to Unit 2

Chapter 5

1 $v = \frac{d}{t} = \frac{24}{6} = 4$; $v = \frac{d}{t} = \frac{33}{5.5} = 6$;

$d = v \times t = 7 \times 17 = 119$;

$d = v \times t = 8.5 \times 11 = 93.5$;

$t = \frac{d}{v} = \frac{160}{8} = 20$; $t = \frac{d}{v} = \frac{420}{14} = 30$

2 $v = \frac{d}{t} = \frac{35}{5} = 7$ m/s

3 $d = v \times t = 8.5 \times 7 = 59.5$ m

4 $v = \frac{d}{t} = \frac{46\,000}{60 \times 60} = 12.8$ m/s

5 $t = \frac{d}{v} = \frac{345}{23} = 15$s

6 $d = v \times t = 15 \times 5 \times 60 = 4500$ m

7 (a) $d = 4 \times 2 = 8$m $v = \frac{d}{t} = \frac{8}{0.5} = 16$ m/s

(b) instantaneous speed.

8 Average speed is 2 m/s; average velocity is zero.

9 (a) Vectors have magnitude and direction. Scalars have only magnitude. Examples are in text.

(b) Distance gives how far you have travelled from start. Displacement is distance in a straight line from start to finish along with direction.

(c) (i) 50 m (ii) 0

(d)

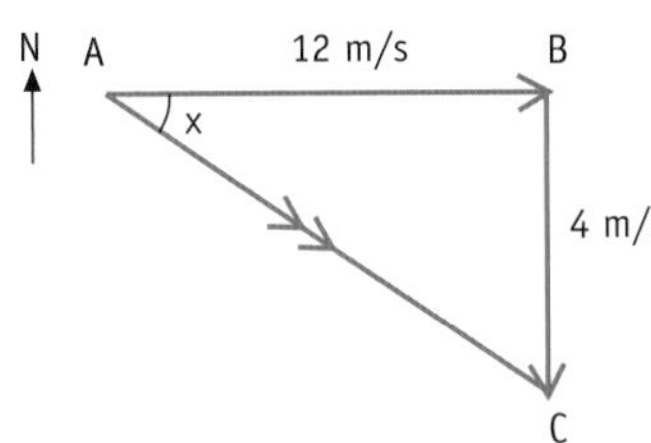

Resultant velocity $= \vec{AC}$

$AC^2 = AB^2 + BC^2 = (12)^2 + (4)^2 = 160$

$AC = \sqrt{160} = 12.6\,\text{m/s}$

$\tan X = \dfrac{BC}{AB} = \dfrac{4}{12} = 0.333$

$x = 18.4^0$

Resultant velocity = 12.6 m/s at 18.4^0 S of E

10

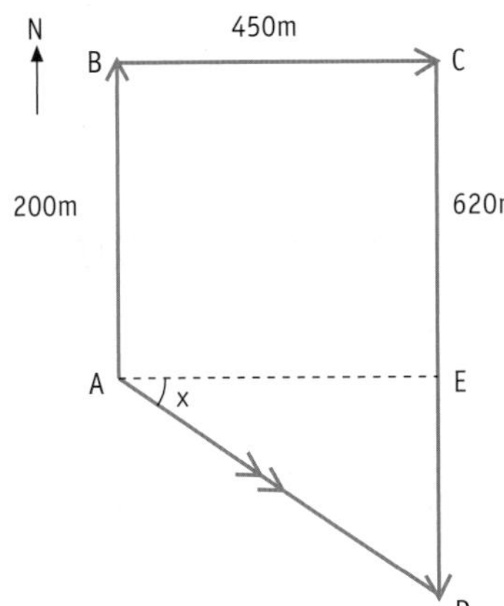

Resultant displacement $= \vec{AD}$

$ED = 420\ \text{m}$

$AD^2 = AE^2 + ED^2 = (450)^2 + (420)^2 = 378\,900$

$AD = \sqrt{378900} = 616\,\text{m}$

$\tan X = \dfrac{ED}{AE} = \dfrac{420}{450} = 0.933$

$X = 43^0$

resultant displacement = 616 m at 43^0 S of E

11

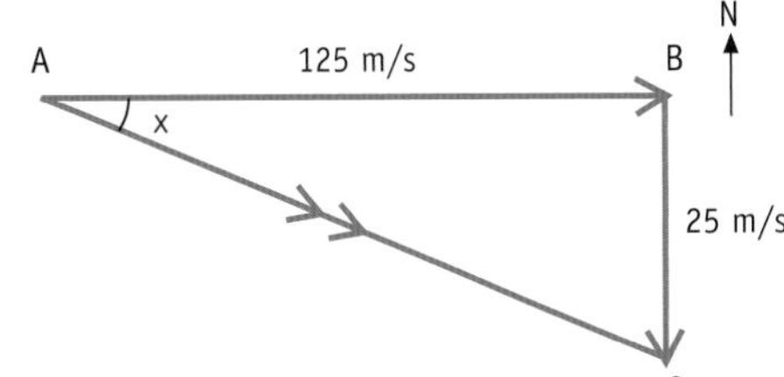

Resultant velocity $= \vec{AC}$

$AC^2 = AB^2 + BC^2 = (125)^2 + (25)^2 = 16\,250$

$AC = \sqrt{16250} = 127\ \text{m/s}$

$\tan X = \dfrac{BC}{AB} = \dfrac{25}{125} = 0.200$

$x = 11.3^0$ Resultant velocity is 127 m/s at 11.3° S of E

12

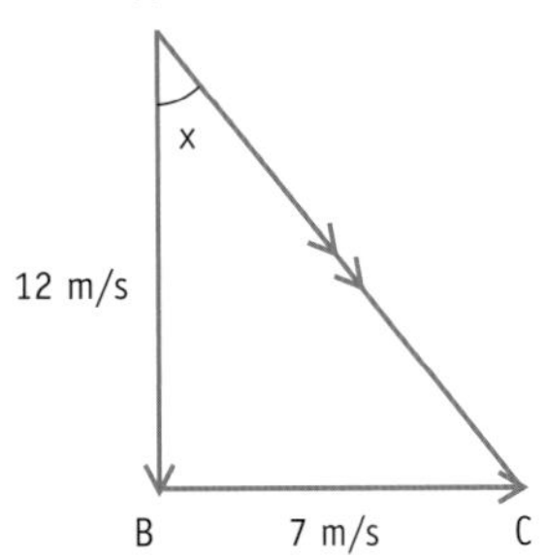

$AC^2 = AB^2 + BC^2 = (12)^2 + (7)^2 = 193$

$AC = \sqrt{193} = 13.9\,\text{m/s}$

$\tan X = \dfrac{BC}{AB} = \dfrac{7}{12} = 0.583$

$x = 30.2^0$

resultant velocity = 13.9 m/s at 30.2^0 E of S

13 5.0; 5.3; 4.4

14 (a) velocity increases by 4 mph every second

(b) less

15 $a = \dfrac{v-u}{t} = \dfrac{5-3}{12} = 0.17\,\text{m/s}^2$

16 $a = \dfrac{v-u}{t} = \dfrac{25-0}{10} = 2.5\,\text{m/s}^2$

$a = \dfrac{v-u}{t} = \dfrac{30-20}{5} = 2\,\text{m/s}^2$

$a = \dfrac{v-u}{t}$

$2 = \dfrac{45-25}{t}$

$t = 10\ \text{s}$

$a = \dfrac{v-u}{t} = \dfrac{14-30}{8} = -2\,\text{m/s}^2$

17 $a = \frac{v-u}{t} = \frac{30-12}{15} = 1.2\ m/s^2$

18 $a = \frac{v-u}{t} = \frac{25-30}{4} = -1.25\ m/s^2$

19 (a) 0-10 s constant acceleration; 10-15 s constant velocity

(b) $a = \frac{v-u}{t} = \frac{8-0}{10} = 0.8\ m/s^2$

(c) area under graph = 80 m

20 (a) 0-150s constant acceleration; 150-250s constant velocity 250–300s constant deceleration

(b) -0.3 m/s^2

(c) 3000 m

21 (a)

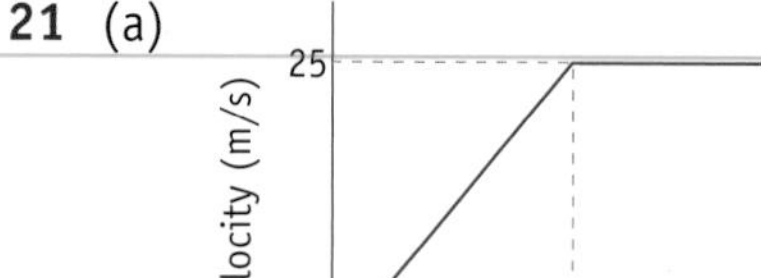

(b) $a = \frac{v-u}{t} = \frac{25-0}{10} = 2.5\ m/s^2$

(c) area under graph =325 m

22 OA constant acceleration; AB constant speed : BC constant deceleration

23 $a = \frac{v-u}{t} = \frac{19-10}{3} = 3\ m/s^2$

24 $a = \frac{v-u}{t} = \frac{8-20}{2.4} = -5\ m/s^2$

25 (a) 2 m/s^2 (b) 75 m

26 2 m/s^2

Chapter 6

1 W = mg
75 = m × 10
m = 7.5 kg

2 100 kg

3 W = mg
= 95 × 1.6
= 152 N

4 W = mg
70 = m × 10
m = 7 kg

5 (a) Frictional force due to air resistance has increased
(b) Put bicycles flat on roof to produce a more streamlined shape

6 No air resistance

7 F_{un} = ma
3000 = 250 × a
a = 12

F_{un} = ma
40 000 = 2000 a
a = 20

F_{un} = ma
50 = m × 2
m = 25

F_{un} = ma
10 000 = m × 5
m = 2000

F_{un} = ma
= 35 × 7
= 245

F_{un} = ma
= 100 × 7.5
= 750

8 F_{un} = ma
130 = m × 3
m = 43 kg

9 (a) F_{un} = ma
= 1250 × 5
= 6250 N
(b) Frictional forces need to be overcome

10 (a) F_{un} = 60 − 12 = 48 N
(b) F_{un} = ma
48 = 12 a
a = 4 m/s^2

11 (a) W = mg
= 350 × 10
= 3500 N
(b) F_{un} = ma
= 350 × 8
=2800 N
(c) thrust = 3500 + 2800 =6300 N

12 (a) Horizontal graph shows a straight line parallel to time axis with value of 15 m/s for 6 s.
Vertical graph straight line from origin going to 60 m/s in 6 s.

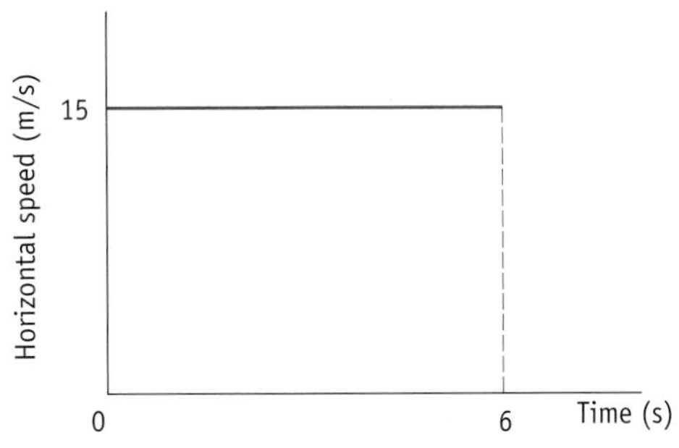

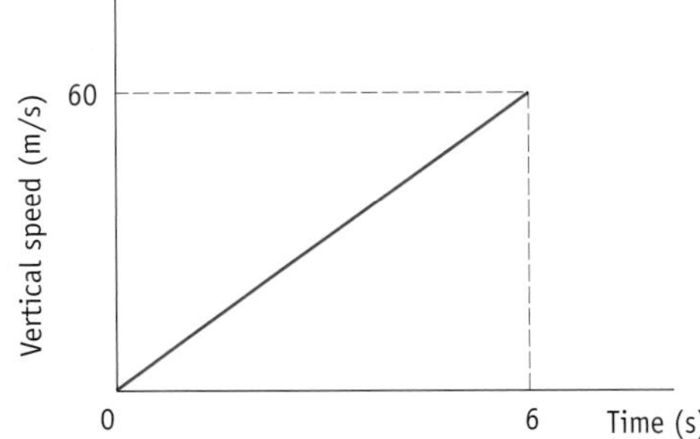

(b) d = v t
= 15 × 6
= 90 m

(c) Area under vertical speed/time graph = ½ × 6 × 60
= 180 m

13 (a) Diagram should show engine force in one direction and air resistance in the other direction.

Air resistance ⟵ ⟶ Engine force

(b) Constant speed means forces are balanced.

14 (a) W = mg
= 60 × 10
= 600 N

(b) W = mg
= 60 × 1.6
= 96 N

(c) W = mg
= 60 × 4
= 240 N

15 F_{un} = ma
= 800 × 3.5
= 2800 N

16 Frictional force

17 (a) Unbalanced force = 200 − 30 = 170 N

(b) F = ma
170 = 25 a
a = 6.8 m/s^2

18 Resultant force

19 26 N/kg

20 (a) d = v × t = 55 × 8 = 440 m

(b) v = u + at
= 0 + 10 × 8
= 80 m/s

Chapter 7

1 Total momentum before = Total momentum after
50 × 8 = (50 + 5) v
v = 7.3 m/s

2 (a) 1000 × 15 = 15 000 kg m/s

(b) Total momentum before = Total momentum after
15000 = (2500 + 1000) v
v = 4.3 m/s

3 Total momentum before = Total momentum after
25 × 5 = (25 + 7) v
v = 3.9 m/s

4 Total momentum before = Total momentum after
1050 × 21 = (1050 + 950) v
v = 11 m/s

5 (a) momentum = m × v = 10 × 6 = 60 kg m/s

(b) Total momentum before = Total momentum after
60 = (20 + 10) v
v = 2 m/s

6 25 × 17 = 425;
350 × 8 = 2800;

$\frac{960}{12}$ = 80;

$\frac{1200}{25}$ = 48;

$\frac{15\,000}{750}$ = 20;

$\frac{800}{20}$ = 40

7 Work done = F × d = 5000 × 2.5 = 12 500 J

8 Work done = F × d = 250 × 3 = 750 J

9 400; 5; 10; 26

10 Work done = F × d = 20 × 30 = 600 J

11 Power = $\frac{\text{work done}}{\text{time taken}} = \frac{10 \times 30}{5} = 60$ W

12 E_p = mgh = 0.2 × 10 × 7 = 14 J

13 (a) E_p = mgh = 50 × 10 × 18 × 0.2 = 1 800 J

(b) Power = $\frac{\text{work done}}{\text{time take}} = \frac{1800}{4} = 450$ W

14 (a) E_K = 0.5 mv² = 0.5 × 0.2 (8)² = 6.4J

(b) 6.4 J

(c) 6.4 = mgh

= 0.2 × 10 × h

h =3.2 m

15 (a) E_p = mgh = 0.5 × 10 × 75 = 375 J

(b) 375 = E_K = 0.5 mv² = 0.5 × 0.5 v²

v = 38.7 m/s

16 (a) E_K = 0.5 mv² = 0.5 × 900 × (15)² = 101 250 J

(b) 101 250 J

(c) Work done = F × d

= F × 200

= 101 250 J

F = 506 N

17 (a) E_p = mgh = 500 × 10 × 8 = 40 000 J

(b) Power = $\frac{\text{work done}}{\text{time}} = \frac{40\,000}{4} = 10\,000$ W

(c) Power changed to heat energy

18 E_K = 0.5 mv² = 0.5 × 20 000 × (1000)² = 1 × 10^{10} J

19 250 J; 15 m/s; 12 kg

20 (a) E_p = mgh = 0.25 × 10 × 3.2 = 8 J

(b) 8 = 0.5 mv² = 0.5 × 0.25 v²

v = 8 m/s

21 (a) power output = change in gravitational E_p/time

= mgh/t

= 8 × 10 × 6/4

= 120 W

Efficiency = $\frac{\text{power output}}{\text{power input}} = \frac{120}{200} \times 100 = 60\%$

(b) Energy changed to heat

22 (a) E_p = mgh = 3 × 10 × 4 = 120 J

(b) E_K = 0.5 mv² = 0.5 × 3 × 8² = 96 J

(c) Efficiency = $\frac{96}{120} \times 100 = 80\%$

Chapter 8

1 (a) It takes 8360 J to change the temperature of 2 kg of water by 1°C.
It takes 2 × 8360 J to change the temperature of 4 kg of water by 1°C.
It takes 16 720 J to change the temperature of 4 kg of water by 1°C.

(b) It takes 16 720 J to change the temperature of 4 kg of water by 1°C.
It takes 5 × 16 720 J to change the temperature of 4 kg of water by 5°C.
It takes 83 600 J to change the temperature of 4 kg of water by 5°C.

(c) It takes 83 600 J to change the temperature of 4 kg of water by 5°C.
It takes 2 × 83 600 J to change the temperature of 8 kg of water by 5°C.
It takes 167 200 J to change the temperature of 8 kg of water by 5°C
It takes 167 200 J to change the temperature of 8 kg of water by 5°C
It takes 2 × 167 200 J to change the temperature of 8 kg of water by 10°C
It takes 334 400 J to change the temperature of 8 kg of water by 5°C

2 (a) E_h = cmΔT = 4180 × 0.5 × (58 − 18) = 83 600 J

(b) E_h = cmΔT = 500 × 0.95 × (198 − 18) = 85 500 J

3 (a) E_h = cmΔT = 4180 × 1.2 × (100 − 18) = 401 280 J

(b) $P = \frac{E}{t} = \frac{401\,280}{180} = 2229$ W

4 (a) P = IV = 4 × 12 = 48 W

$P = \frac{E}{t}$

$48 = \frac{E}{(5 \times 60)}$

E = 48 × 300 = 14 400 J

(b) $E_h = cm\Delta T$

$14\,400 = 386 \times 1 \times \Delta T$

$\Delta T = \frac{14\,400}{386} = 37.3°C$

(c) Some of the heat supplied by the heater will be transferred to the surroundings and not to the copper.

5 (a) $E = P \times t = 2000 \times 40 = 80\,000$ J

(b) $\Delta T = \frac{E_h}{c \times m} = \frac{80\,000}{4180 \times 0.4} = 47.8°C$

Final temperature $= 47.8 + 18$
$= 65.8°C$

6 $E_h = ml = 0.8 \times 3.34 \times 10^5 = 2.67 \times 10^5$ J

7 $E_h = ml = 0.2 \times 2.26 \times 10^6 = 4.52 \times 10^5$ J

8 $P = \frac{E}{t}$

$2000 = \frac{E}{80}$

$E = 2000 \times 80 = 1.6 \times 10^5$ J

$E_h = ml$

$1.6 \times 10^5 = m \times 2.26 \times 10^6$

$m = \frac{1.6 \times 10^5}{2.26 \times 10^6} = 0.071$ kg

Unit 2 Exam Questions

1 (a) (i)

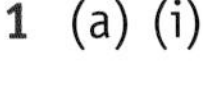

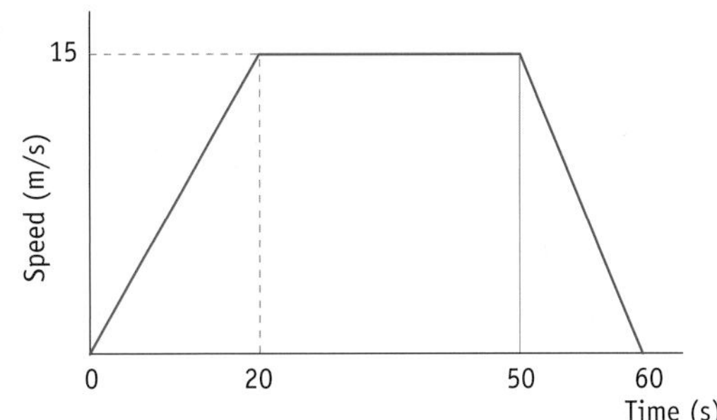

(ii) $a = \frac{v-u}{t} = \frac{15-0}{20} = \frac{15}{20} = 0.75\ m/s^2$

(iii) $a = \frac{v-u}{t} = \frac{0-15}{10} = \frac{-15}{20} = -1.5\ m/s^2$

(iv) distance travelled from 50 s to 60 s = area under graph
$= ½ \times 10 \times 15$
$= 75$ m

(b)

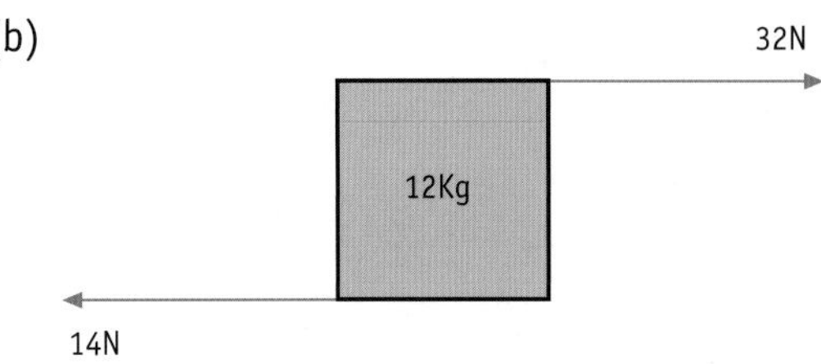

(i) $F_{un} = 32-14 = 18$ N

$a = \frac{F_{un}}{m} = \frac{18}{12} = 1.5\ m/s^2$

(ii) $v - u = at$

$v - 0 = 1.5 \times 8$

$v = 12$ m/s

(iii) Acceleration will decrease – bigger force of friction now present so smaller unbalanced force.

2

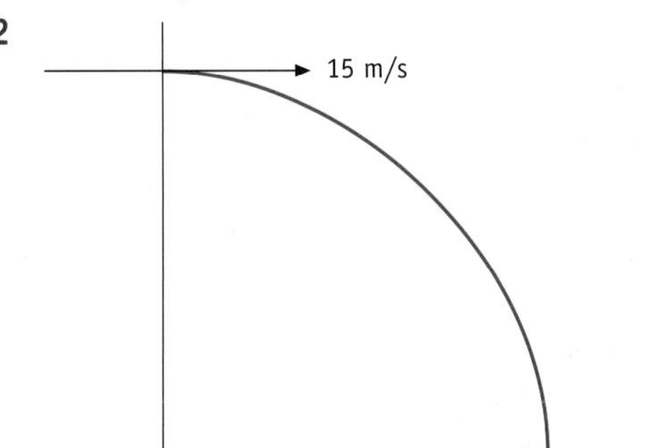

(a) $d = v \times t = 15 \times 3.5 = 52.5$ m

(b) $v - u = at$

$v - 0 = 10 \times 3.5$

$v = 35$ m/s

(c) (i) horizontal distance is greater.

(ii) vertical speed still the same since the ball will be in the air for the same length of time as before.

3 (a) total momentum before collision is equal to the total momentum after collision as long as no external forces act.

(b) total momentum before $= 1000 \times 15$
$= 15\,000$ kg m/s

total momentum after $= (1000 + 1200) \times v$
$= 2200$ v

$2200v = 15\,000$

$v = 6.8$ m/s

(c) Velocity needs a size and direction to describe it fully.

(d) During the collision, according to Newton's First Law, the passenger will continue to move forward at 15 m/s. From Newton's Second Law an unbalanced force must be applied to the passenger to stop this. The seat belt applies the unbalanced force to the passenger and stops them moving forward.

4 Water at 20°C to 0°C, $E_h = cm\Delta T$
$E_h = 4180 \times 0.15 \times 20$
$E_h = 12\,540$ J
Water at 0°C changes into ice at 0°C, $E_h = ml$
$E_h = 0.15 \times 3.34 \times 10^5$
$E_h = 50\,100$ J
Ice at 0°C to –19°C, $E_h = cm\Delta T$
$E_h = 2100 \times .0.15 \times 19$
$E_h = 5985$ J
Total energy removed = 12 540 + 50 100 + 5985
= 68 625
= 68.6 kJ

5 (a) $E_h = cm\Delta T$
$E_h = 4180 \times 1.2 \times 12.5$
$E_h = 62\ 700$ J

(b) $P = \frac{E}{t} = \frac{62\,700}{23 \times 60} = 45$ W

Assuming that all the energy supplied by the heater is absorbed by the water.

(c) $P = IV$
$45 = I \times 12$
$I = \frac{45}{12} = 3.75$ A

6 (a) 80°C
(b) $Eh = cm\Delta T$
$P \times t = cm\Delta T$
$100 \times 330 = c \times 0.4 \times (80 - 20)$
$c = \frac{33\,000}{24} = 1375$ J/kg °C

(c) $E_h = ml$
$P \times t = ml$
$100 \times (830 - 330) = 0.4 \times l$
$l = \frac{50\ 000}{0.4} = 125\,000$ J/kg

Answers to Unit 3

Chapter 9

1 $v = \frac{d}{t} = \frac{3}{0.01} = 300$ m/s

2 $d = v \times t = 330 \times 3.3 = 1089$ m (6.6 s is the time to go to the mountain and back so the time to the mountain is 3.3 s)

3 $d = v \times t = 1500 \times 0.00002 = 0.03$ m

4 energy

5 (a) 1.5 m (b) $3 \times \lambda = 18$ so $\lambda = 6$ m (c) $f = \frac{125}{10}$ = 12.5 Hz
(d) $v = f \times \lambda = 12.5 \times 6 = 75$ m/s

6 1.8; 340; 2×10^{-6}; 6; 133; 5×10^{14}

7 $v = f \times \lambda = 5 \times 4 = 20$ m/s

8 $v = f \times \lambda$
$1500 = 2 \times 10^{6}\lambda$
$\lambda = 7.5 \times 10^{-4}$m

9 $f = \frac{v}{\lambda} = \frac{600}{0.025} = 24\,000$ Hz

10 3×10^8 m/s

11 (a) Particles vibrate at right angles to the direction of energy transfer.
(b) Any radiation in the electromagnetic spectrum

12 Particles vibrate parallel to the direction of energy transfer

13 Gamma, ultraviolet, visible light, microwaves, TV

14 (a) infrared (b) infrared (c) Gamma or X-rays

Chapter 10

1 (a) A thin piece of flexible glass that transmits light within it

(b)

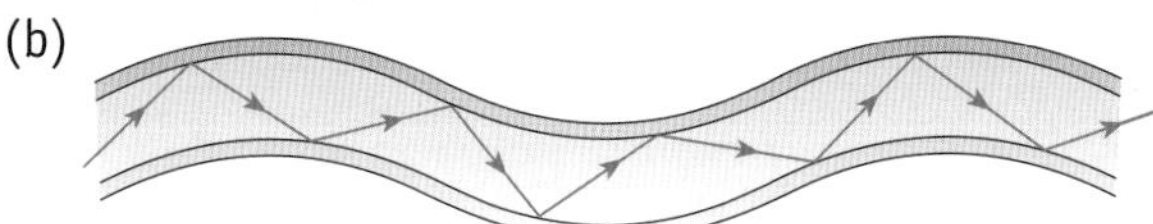

2 (a) and (b) As the diagram in Figure 10.1 page 122.

(c) 60°

3 (a) Rays travel in a straight line as a parallel beam

(b) Focus

4

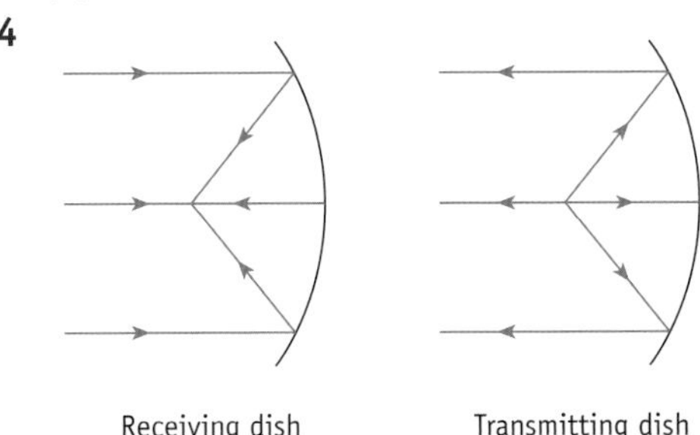

One to transmit and the other to receive the signals.

5

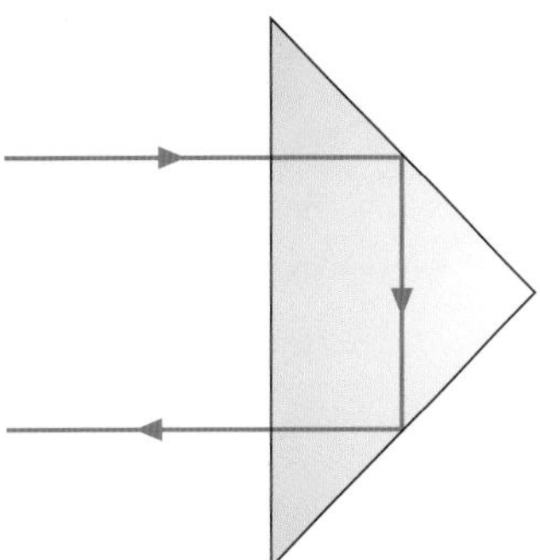

The angle of incidence in the glass is greater than the critical angle. The ray is totally internally reflected in the glass.

6 (a) Light

(b) See pages 125 to 127

(c) Cost less, less need for repeaters, reduced interference, greater signal capacity.

Chapter 11

1 (a) Refraction

(b) See text on page 136.

(c) Angle of incidence is greater than the angle of refraction

2

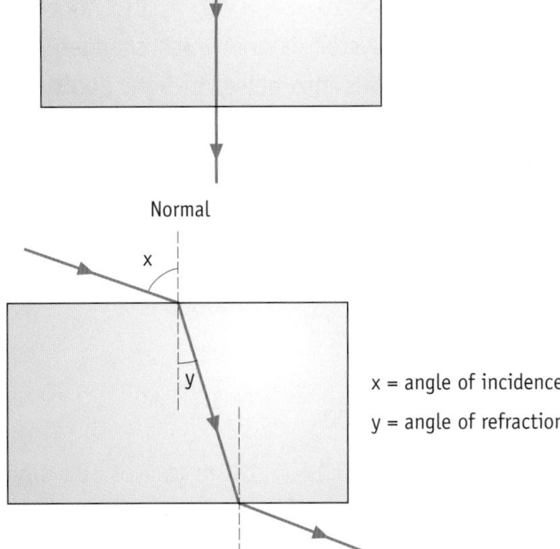

3 (a)

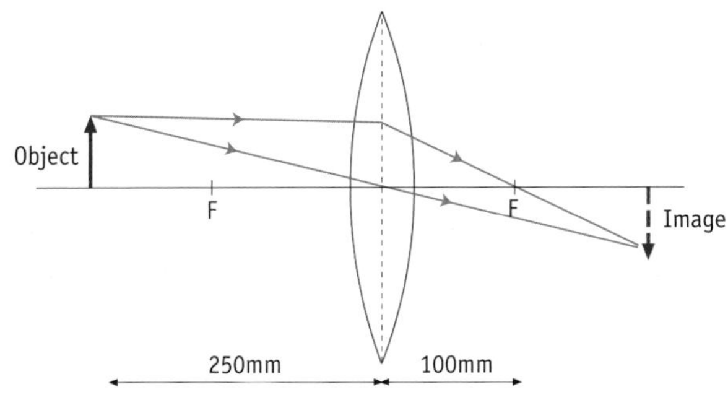

(b) length of image = 8 mm

length of object = 10 mm

image is 0.8 times smaller.

4 (a) Less than the focal length

(b) (i)

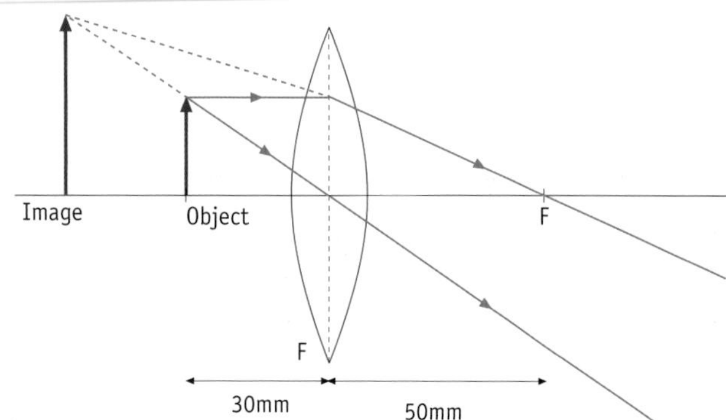

(ii) length of image = 25 mm
length of object = 10 mm
image is 2.5 times bigger than object.

5

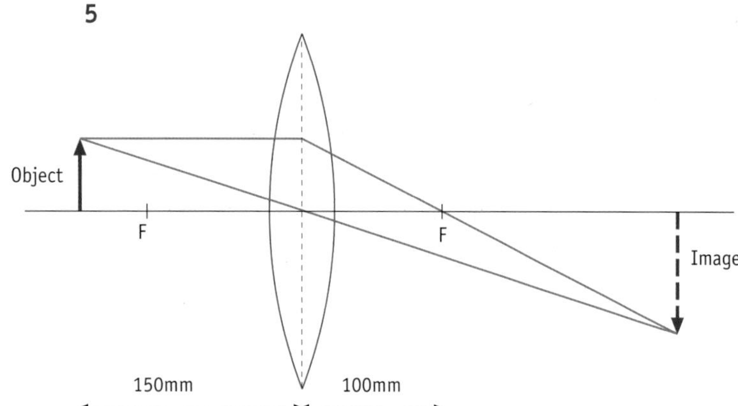

6 See pages 139 and 140.

7 (a) long sight, since rays of light from a close up object are focused beyond the retina and so a convex lens is needed to refract the rays of light more so that they focus on the retina.

(b) $\text{power} = \dfrac{1}{\text{focal length}}$

$2.5 = \dfrac{1}{\text{focal length}}$

focal length = 0.4 m

8 10; −4; 250 convex; 200 concave; −3.3; 2.5

9 (a) short sight (b) concave lens

(c) $\text{power} = \dfrac{1}{\text{focal length}}$

$= \dfrac{1}{0.6}$

= −1.7 D concave lens needed to correct sight defect

10 (a) (i) Y and Z – both are convex lenses and this is the lens type needed to correct long sight.

(ii) X – a concave lens and this is the lens type needed to correct short sight.

(b) Thinner and not scratched unlike conventional spectacle lenses.

Unit 3 Exam Questions

1 (a) (i)

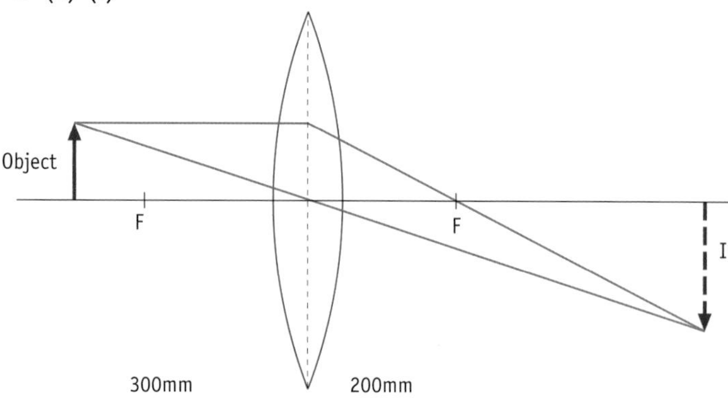

Image is formed on opposite side of lens from object as shown.

(ii) real and magnified

(b) (i) long sight

(ii) $\text{power} = \dfrac{1}{\text{focal length}}$

$= \dfrac{1}{0.2}$

= 5D

2 (a) 27 MHz = 27,000,000 Hz. Frequency is the number of waves produced in one second – in this case 27,000,000 waves are produced in one second

(ii) $\lambda = v/f = 3 \times 10^8/27 \times 10^6 = 11.1$ m

(b) (i) A fibre optic system is one which changes electrical signals into pulses of light and then sends them down an optical fibre using total internal reflection and at the end of the fibre the light signals are changed into electrical signals.

(ii) $t = \dfrac{d}{v} = \dfrac{20\ 000}{2 \times 10^8} = 0.0002$ s

3 (a) $\lambda = \dfrac{v}{f} = \dfrac{1500}{5\ 000\ 000} = 3 \times 10^{-4}$ m

(b) d = v t = 1500 × 0.12 = 180 m (t = 0.12 s since this is the time for the signal to reach the shoal)

4 (a) 25° (b) 40°

(c) +3.5 D – more refraction required to focus light from a near object on the retina so greater power of lens

Answers to Unit 4

Chapter 12

1 (a) Proton positive charge; neutron no charge; electron negative charge
(b) proton and neutron in nucleus, electron surrounding nucleus

2 (a) (i) helium nucleus (ii) 50 mm
(b) Gain or loss of an electron to produce a charged particle
(c) High ionisation density so has more effect on the cells

3 (a) see page 156
(b) Gamma is the only radiation which will pass through tissue and so reach the detector which is outside the body. The other radiations are absorbed by the body and so cannot escape and reach the detector.

4 (a) A fast moving electron
(b) Aluminium

5 Any use of radiotherapy see page 155

6 (a) gamma (b) (i) As close to the tumour (ii) high ionisation density so alpha particles are absorbed by the tissue and may not reach the tumour unless the source is close to it.

Chapter 13

1 (a) Number of nuclei which disintegrate per second.
(b) Bq.

2 400; 20 000; 5×10^{-3}; 30; 6×10^{8}; 4.5×10^{6}

3 2.5; 4; 5×10^{-4}; 0.315; 0.2; 50

4 The absorbed dose; the type of radiation; the type of tissue.

5 $H = D\,\mathbf{w}_R$ and measured in sieverts (Sv)

6 $H = D\,\mathbf{w}_R$
$= 6 \times 10^{-4} \times 10$
$= 0.006$ Sv

7 0.001; 0.06; 0.0001; 0.01; 2; 20

8 See pages 165 to 167 for a list of factors.

Chapter 14

1 Time taken for half the radioactive nuclei to disintegrate is 6 hours

2

Activity (kBq)	Time (days)
640	0
320	8
160	16
80	24
40	32
20	40

Activity after 40 days is 20 kBq.

3

Activity (kBq)	Time (hours)
600	0
300	half life
150	half life
75	half life

3 × half life = 24 hours

$$\text{Half life} = \frac{24}{3} = 8 \text{ hours}$$

4 36 hours is 6 half lives. Half 50 kBq 6 times and activity is 781 Bq

5

Activity (kBq)	Time (days)
1	24
2	16
4	8
5	0

Initial activity = 8 kBq

6 See page 171

7 Increase distance away from source; put shielding between you and the source.

8 In hospitals; school storerooms; nuclear power plants etc.

Chapter 15

1 Advantages - no gases produced to damage environment; lots of material available for a long time; Disadvantages – storing radioactive waste and the dangers of accidents.

2 Nucleus will split into smaller pieces and release more neutrons and some energy. These neutrons will hit other nuclei causing the process to repeat.

3 To absorb neutrons so that the fission reaction is controlled. Boron is a good absorber of neutrons.

4 See pages 181 to 182

5 See pages 182 to 184

6 (a) total mass of fuel rods = 1700 × 14 = 23 800 kg

mass of uranium = 3% of 23 800 = $\frac{3}{100}$ × 23 800 = 714 kg

(b) mass of uranium = 714 kg

Energy released = 714 × 1 × 10^{14} = 7.14 × 10^{16} J

Unit 4 Exam Questions

1 (a) Time taken for half the radioactive nuclei to disintegrate

(b)

Activity (Bq)	Time (years)
1 000 000	0
500 000	1
250 000	2
125 000	3
62 500	4
31 250	5
15 625	6

Activity after 6 years is 15 625 Bq

(c) The absorbed dose; the type of radiation; the type of tissue

(d) (i) $D = \frac{E}{m} = \frac{0.35}{70} = 0.005$ Gy

(ii) $H = D \times w_R = 0.005 \times 20 = 0.1$Sv

2 (a) see page 180

(b) see pages 178 and 179

(c) see pages 180 to 182

(d) see pages 182 to 184

3 (a) gain or loss of an electron to produce a charged particle

(b) alpha

(c) (i) Only gamma will pass from the patient through body tissue to be detected outside

(ii) A longer half life could be dangerous to the patient as they would be exposed to high level of radiation for a long time and a shorter one would not allow the radiation to reach the organ and be detected.